GREAT JOBS

FOR

Math Majors

Stephen E. Lambert and Ruth J. DeCotis

McGraw·Hill

New York   Chicago   San Francisco   Lisbon   London   Madrid   Mexico City
Milan   New Delhi   San Juan   Seoul   Singapore   Sydney   Toronto

The **McGraw·Hill** Companies

Library of Congress Cataloging-in-Publication Data

Lambert, Stephen E.
    Great jobs for math majors / by Stephen E. Lambert, Ruth J. DeCotis. — 2nd ed.
      p.   cm.
    Includes index.
    ISBN 0-07-144859-4 (pbk. : acid-free paper)
    1. Mathematics—Vocational guidance.   I. DeCotis, Ruth J.   II. Title.

  QA10.5.L36   2006
  510′.23—dc22   2005047956

1 2 3 4 5 6 7 8 9 0 DOC/DOC 0 9 8 7 6 5

ISBN 0-07-144859-4

McGraw-Hill books are available at special quantity discounts to use as premiums and sales promotions, or for use in corporate training programs. For more information, please write to the Director of Special Sales, Professional Publishing, McGraw-Hill, Two Penn Plaza, New York, NY 10121-2298. Or contact your local bookstore.

This book is printed on acid-free paper.

This book is dedicated with great affection to my CAGS colleagues:
Jack Barry, Gail Carr, Alan Davis, Maria Dreyer, Daniel Ferrera,
Kathleen Norris, and Marianne True. During the writing of this book,
I was uplifted and supported by their humor and intelligence and by
their remarkable individual gifts. I cannot believe my good fortune
in knowing each of you.

—Stephen E. Lambert

This book is joyously dedicated to my husband, Terry, for loving me
through thick and thin, no matter what I was trying to do to balance my
life. To my wonderful children, Greg, Curtis, and Erin, and their families;
to my father, David, and my mother, Olive; to my seven sisters; and to
my friends Lee and Nancy. All of them were with me in spirit, listening
to my joys and concerns. All of you supported me as I balanced my
commitments on an overbooked calendar of my life's events, which
included taking time out to coauthor my first book with Stephen.

—Ruth J. DeCotis

Contents

Acknowledgments

The authors want to acknowledge the superb contributions of Kendra Cantin, our Undergraduate Fellow in Career Services for 1996–1998. Her research efforts, telephone skills, and computer facility added immeasurably to *Great Jobs for Math Majors*. Beyond those stellar efforts, she has been a wonderful colleague, whom we have missed tremendously since her graduation. Thank you, Kendra, from both of us. The editors wish to thank Barb Donner for revising this second edition.

Introduction

Like the Roman god Janus, mathematics has two faces. One face is the abstract world of numbers and the intellectual challenges that hold mathematicians spellbound. Separated from human values and mundane concerns, abstract mathematical reasoning can lead mathematicians to deep and powerful conclusions that ultimately influence other branches of thinking. A good example of this spreading-ripples effect is the fascination with a school of management theory that has been based on a rejection of Newtonian physics and embraces a quantum perspective on how organizations shape and reshape themselves.

Philosophy asks many questions of us, and two of the most profound are What is truth? and How can we know it? Though we are still wrestling with these questions, the science of mathematics has been one of the primary sources of the truths that humanity has discovered. Certainly, to many people, math has its place in history as the most challenging example of the powers of the human mind.

For even a modestly educated person, it is not difficult to recognize the importance of the work of the chemist dealing with molecules, the biologist with DNA, or the physicist with the black holes of space. To approach some of the arcane heights of mathematics, however, not even analogies or metaphors can help us appreciate, much less understand, the heights achieved by the great masters of abstract mathematics.

Mathematics holds its place in great art, also, not only in architecture and music but in the search for beauty itself. The study of fractals, for example, has given us some insight into why we are attracted to and utilize certain motifs—the scroll, for example—over and over again in art. Designs such as the scroll and others have pronounced similarities to certain fractal patterns.

The second face of mathematics, and the one with which this book deals, is the practical face. Though most people are not at all familiar with abstract mathematics, the more accessible and familiar science of practical mathematics is seldom called into question by anyone, including those people with no mathematical skills whatsoever. Most people know the value of math skills in everyday life, and many simply wish they had better skills in math than they do.

In spite of a math envy that exists in the general public, however, math majors themselves are often unaware of the tremendous practical value of their own degrees—and they may become acutely aware of this lack as they approach graduation. We hope, as you read this book, that you will be surprised at how valuable your education has been. More important, we hope the material in this book will help show you how you can combine who you are and what you want with the education that you have achieved to find a spot in the employment market that's just right for you.

As career counselors, it's no surprise to us that graduating math majors can be almost completely unaware of the wonderful possibilities that exist for them. After all, math majors have been involved in a very demanding curriculum. There is seldom time in an advanced math course to cover all the material required, much less to explore the real-world applications for career potential as well.

Nor are the demands of delivering the content of the courses the only reason why careers are not a big topic of discussion in math classes. Another reason is that many math faculty members have had a fairly direct and linear progression through college and graduate work to their current positions. Though they may have had some other kinds of employment using their math skills and may also be involved in research on math topics, their knowledge of the scope of career possibilities may be somewhat limited. Mathematics is not an isolated body of knowledge and, to be most useful, should be taught in contexts that are meaningful and relevant to learners. If, however, after taking a math course, you could not answer the question "Now, what am I going to do with this?" your instruction was probably not unusual but simply lacking in how to apply what you learned to real-life situations.

It can be very discouraging to reach graduation, no matter what your major, and find a big gap between what you have learned and the reality of an actual job search. Add to that the fact that most people know only a few sources of information about the employment possibilities that exist for job applicants with their educational preparation. Advice from parents, friends, or teachers is helpful, but it cannot take the place of your own personally designed and structured exploration of the various career pathways available to you with your degree. We hope to help you with that organized exploration in this book.

Mention that you are a math major to most people, and they respond by saying, "Oh, I'm terrible at math," or "I wish I were smart enough to do math." Many people are very insecure about their math skills. You know this to be true because you've already had some of these conversations and experienced these reactions. It is a frequent puzzle to math graduates that so many people are put off by this subject because of a bad experience in school or perhaps by the scary look of written math. But now, as you approach the job search, you may find that you're suddenly grateful for the public's admiration of people who can "do math." It may help you get a job!

The mathematics major is a wonderful degree, because there are so many—almost unlimited—opportunities for graduates who have superior math skills and some openness and curiosity about a variety of possible occupational settings. As Chapter 6 makes clear, women and minorities are still not well represented in math careers, so employers usually make a special effort to encourage members of these groups to apply. Jobs requiring significant educational preparation in mathematics are always among the top fifty jobs listed in the *Jobs Rated Almanac*.

You can be confident that public esteem is high for your degree and there is usually sufficient demand for employment of math majors. To have the most flexibility in your job search, your academic record in your major should be superior—not just your academic record as a paper document, but your acquisition of the skills, reasoning ability, and problem-solving techniques that those high grades should accurately represent.

In addition, you will notice that this book emphasizes your communication skills as well. The reason we have done this is that the ability of a math major to communicate effectively is a unifying element for all the jobs discussed. You will notice in job after job, as presented in this book, you are called on to work in teams and groups, to explain and teach, to clarify and listen. Application of mathematical models and use of math techniques to analyze and solve problems come only after problems have been discussed and understood. All of this interaction and interpersonal work demand excellent oral and written communication skills.

You know that mathematics is, at its heart, a linguistic activity, with the ultimate potential of precise communication of meaning. In math, we also have the satisfaction and resolution of proofs. When Pythagoras discovered the Pythagorean theorem, he knew with certainty that the theorem could be communicated exactly to mathematicians everywhere.

The clarity of communication afforded by mathematics has been dramatized in a number of movies about outer space. In *Close Encounters of the Third Kind* and a more recent movie, *Contact*, math is the language of choice between us and our visitors! In *Contact*, Jodie Foster plays an astronomer who is searching the heavens for extraterrestrial life signs. When contact is made, it is made through a pattern of binary numbers. The language of numbers is certainly a brilliant human concept, and, as this fictional movie demonstrates, math holds the possibility of being a more effective bridge of meaning than vernacular language. What a great cultural lineage you help to perpetuate when you bring your math skills into the workplace.

We know that math is more than a skill. Mathematics embraces an entire world of ideas, and the perpetrators of its history have been some of the most brilliant minds ever encountered, including Pythagoras, Euclid, Descartes, and Srinivasa Ramanujan, the genius mathematician who came to Cambridge University in England from a humble village in India to electrify the thinking of the world about numerical theory.

This book is not a history of math, of course, but a guide for college math graduates who need an answer to the question "How can I use my math degree to forge a productive and happy life for myself?" To answer that question, we have written a self-contained guide that will not only help to prepare you for the job search in general but will direct your thinking about mathematics careers into a number of possible pathways. This is not generalized how-to advice but very specific, concrete information on different industries and job areas within them. For each functional area we discuss, we look at definition of the career path, working conditions, training and qualifications, earnings, and strategies for finding these jobs. The following is a quick look at each of the chapters and what you will find there.

Job Search Preparation

In Chapter 1, "The Self-Assessment," perhaps the most important aspect of the job search is addressed. We ask you to take an inventory of who you are, how you like to work, and what you want to do in the workplace. Reflecting on your personality, your work experience, and your educational background will prepare you to write and talk about your plans to others.

If you want to act on the suggestions in this book, you need a résumé and the ability to write a good cover letter. You're not really ready to job hunt until you have both of these documents ready to go. Chapter 2, "The Résumé and Cover Letter," teaches you what you need to know and provides some practical examples particularly designed for math majors.

No matter which direction your job search is taking, there's good advice in Chapter 3, "Researching Careers and Networking," regarding where to learn more about the career you've chosen. We discuss the value of researching the job setting, the best job directories, and various sources of job information. When you add this information to the specific professional associations and websites listed at the ends of Chapters 6, 7, 8, and 9, you will have your own key to developing a personal job search library.

Chapter 3 also gives you solid techniques for networking successfully. Every successful job seeker knows that you shouldn't keep your job search a secret! Tell your friends, teachers, and past employers, and have your parents tell their friends. The more people you talk to, the greater the possibility you'll connect with someone who may have some interesting information or a contact who may prove helpful. We also teach you how to leave the network you've established once you've been hired, in case you ever need to reactivate it at some point in the future.

Chapter 4, "Interviewing and Job Offer Considerations," gives you a good understanding of what the employer hopes to accomplish during an interview and how you can make the interview work for you. It's an important chapter. The best résumé, the sharpest cover letter, and all the networking in the world won't overcome a bad interview. One reason interviews sometimes don't work is because we don't understand the concept behind them and what our role is.

The chapter also addresses some details job seekers often forget. We discuss what job search activities take place after the interview and how your post-interview behavior can enhance or detract from your chances for success. Once you've been offered a job and before you sign on the dotted line, we ask you to consider a few points by giving you some tips on negotiation strategies and how to compare offers.

The Career Paths

Before you read Chapters 6 through 9, in Chapter 5, "Introduction to the Mathematics Career Paths," we give you a broad overview of each of the career path chapters and some advice on how to approach comparing and contrasting the various jobs.

Teaching any subject is an art, and Chapter 6, "The Math Job You Know Best," attempts to convey some of the special demands of the teaching profession. Whether it's at the elementary school or college level, teaching will draw on far more than your mastery of mathematics. We give you a very realistic overview of the profession at every grade level.

In Chapter 7, "Jobs Using Math as a Primary Skill," the "classic" math graduate jobs are discussed—but with a twist. Actuary, mathematician, statistician, and operations research analyst positions are discussed, with special attention paid to those job factors (training, salary, and career outlook) that will be important in your decision making.

Chapter 8, "Working Toward an Advanced Degree," looks at market and financial analysts, and research analyst and associate positions, which are available to both math and other quantitatively skilled graduates. This chapter addresses these jobs both as destination career positions or as possible shorter-stay jobs if you are contemplating a return to graduate school.

Chapter 9, "Math in the Marketplace," focuses on jobs where your math skill is an invaluable resource. Each of the career paths of buyer, sales representative, and purchasing agent welcomes a variety of majors but especially values the math major. Since math is not a primary skill for these positions, we address the other important qualifications to help you consider whether your career may be in a job buying and/or selling in the workplace.

Each of the individual career paths is distinct, but there are some unifying themes. Applications, problem solving, and reasoning are some of the elements that bind together most math jobs. All the jobs in this book share a connection with these three skills. Of course, problem solving and reasoning have been the mainstays of mathematics instruction since the beginning of teaching, and applications have recently been through a major revival of interest. Data analysis projects or opportunities for creating production and financial models occur in settings as diverse as the popular music industry, fashion design houses, major medical facilities, and the automobile industry. There is no reason why you can't take your skills and interests and apply math methodologies and concepts in a job setting of your choice.

We hope you recognize as you read these chapters not only the emphasis on communication but our general strong feeling that one of your principal roles in the workplace will be to communicate the duality of mathematics, that is, its usefulness and its beauty. In many ways, computers have, with their ever-expanding software applications and speed of performance, unfortunately served to further distance some people from approaching and understanding mathematics. You can help bridge this gap by utilizing all the skills in your repertoire, including models, pictures, diagrams, and graphs, to propose the useful possibilities of math in business situations. This approach will help coworkers who are math phobic to deal with abstract ideas by having a visual realization of them. Of course, none of your tools needs to be static, because you can use the same technology that has alienated some people to bring others closer to math. These tools don't have to create distance because, when used by the skilled mathematician, they become powerful tools for communication. Mathematics is changing, and your role is changing as a mathematician.

For example, in recent years, the mathematics field has increasingly recognized the disservice done to young women during their school years when interest and achievement in math were encouraged more in young men. That issue is now being addressed across the country, and the growth we'll see of women in math professions will be remarkable and far reaching. Regardless of your own gender, it is hoped that in your math career you will provide encouragement and positive role modeling for all young people.

The four career paths outlined in this book display the virtuosity of math and the fact that math is present and has penetrated every aspect of business and industry. Math takes on issues and problems of quantity, space, pattern, arrangement, structure, and logic that help us define, understand, and build our world. Each career path contains suggestions for countless other possibilities where your math degree will add an essential ingredient to the success of an organization.

Another exciting aspect about the variety of jobs available to the math major will be of personal importance to you. The fact that math majors can be employed in a variety of settings increases your chances of finding a job that both utilizes your educational preparation and satisfies your own personal wishes and needs. Do you want the fast-paced life of a major city, or would you prefer to work in a rural environment? With many analyst jobs and consulting positions, the city will be your home. As a math teacher, on the other hand, you could settle almost anywhere.

Do you prefer a socially interactive workplace, or do you like to work alone and uninterrupted for long stretches of time? You can work in international banking or in the machine tool industry. You could travel year-round or have a desk job. There are jobs that pay very well for math majors and provide the opportunity to earn very high incomes. You'll find all of these issues addressed in the exploration of the career paths that follow.

Learning about these options should give you another reason to turn more than once to Chapter 1 and refine your list of work values—those factors that you consider important or essential in a job. Your list may be a combination of financial reward, stability or risk, potential for growth or a chance to learn new skills, or any combination of these and other values. Whatever the special values included in your list, our belief is that there is a job that uses your math background and can satisfactorily meet most of your other needs and demands as well.

Use an Internship to Test the Waters

If you're feeling uncertain about the kind of job that might be best for you, you might want to consider a student internship. There is no question that an internship is easier said than done. The fact that many are unpaid presents real challenges with the high cost of college today. Many students feel that they cannot afford to work without pay during the summer or other breaks from school.

Important considerations, however, include the following. Giving up some income now can pay real dividends at graduation time. Many internships turn into real jobs for graduates. For others, the internship provides deeper, more concrete insights into their own skills and an appreciation of the roles they can play in an organization. An internship will fill your basic need to learn more about the workplace and where you might fit in. Other benefits also result, such as finding it easier to connect with people when you're interviewing, having more to say, and, as a direct result of your internship, more to offer so you'll stand a better chance of getting hired.

Internships often require relocation. If you live or go to college in a non-metropolitan area, nearby internships will probably be scarce. The greatest concentration of internships tends to be in and around metropolitan areas where there are more organizations that can afford to train temporary, non-paid additions to their staff and provide meaningful work for them. You may have to seek the hospitality of a relative or one of your parents' friends or seek inexpensive temporary housing in order to put yourself where the internships are located.

Finding and applying for internships are fair-sized tasks, though not as difficult as finding your first job. In fact, every activity you'll do in searching for and obtaining your internship is similar to your actual career job search.

Just as with full-time jobs, not all good internships are published in directories or periodicals. Some internships are described on flyers or in brochures that are received in the math department or career office. Check directories and bulletin boards in both places, as well as at your college library. In our career services office, we use the Internet for up-to-date internship listings. A simple search for "Math Internships" will give you several pages of a variety of sources, such as government lists, various university and college lists, insurance and actuary lists, American and European lists, and many more.

Some organizations provide for information requests and applications by e-mail; others require application by mail. Expect to write many cover letters. Send them out with your résumé, and be sure to do it early. Many summer internships have January or February application deadlines. Each year in our career office, we assist scores of students with their internship materials. Invariably those students who did most of their mailings around Thanksgiving vacation seem to do the best in securing the sites they want.

Expect to receive requests for transcripts (both official and unofficial), essays of intent or goals, or to complete formal applications that may include essay writing. All of this correspondence must be perfect, because in most cases, interns are chosen on the basis of the written materials submitted as well as phone interviews. Your written work constitutes your first impression on the people who will be selecting the interns.

Mirroring the competitiveness of the job market in general, the competitiveness for internships today provides a good practice environment for learning how to apply, interview, follow up, accept, and begin a new job. You can expect a well-chosen internship experience to stay with you for the rest of your life. That summer or term in a government agency, actuarial office, corporate research and analysis department, or with an employer in industry will become part of your experiential base as long as you are in the workforce. In innumerable ways, an internship represents a smart move for your future.

By now, you understand the responsibility you have to make the most of the rest of your college years, and you have some specific techniques by which you can do that. You've heard some people say that college isn't the real world and you can just float along, "treading water." We guarantee that if you get out and meet some recent alumni of your school, they will assure you that college has the potential for being as real a world as you want to make it and every bit of experience you can gain is an important aspect of your preparation for a career in mathematics.

Perhaps you are reading this book as a junior or even a senior approaching graduation. Your most important goal right now is probably determining a strategy for finding that job. How do you go about that, considering all you've learned about the changing climate in the employment market?

Realize that You Will Have Many Careers and Plan for That

You've seen all the warning signs that regardless of how talented you are shifts in employers and employee bases do occur and some people are forced to move on. Those who have the most successful transitions have prepared themselves and usually do most of the following things.

Watch trends. If, at some future date, you are in a statistician job and you see an increasing use of certain new software programs, it's a clear sign not only that you need to be acquiring those new skills for the future but that the employer who produces such software may also be a new employment site. Stay abreast of changes in your field by reading professional journals and association newsletters so you have some ready reference points if you do need to jump-start your career at some point. At the end of each career path chapter, we give you names, mail addresses, e-mail addresses, and websites for many of the top professional associations in those fields. It's not too early to join, and many have attractive membership prices for students.

Understand your needs. Every time you make a job change, you have to address issues such as geographic location, housing options, pay levels, your economic and consumer patterns, the duties and responsibilities of the new position, your interest and belief in the nature of the work, and a thousand other issues that profoundly impact who you are and how you live and work. This is a good point in your life to begin to appreciate and understand more consciously and intentionally what you need and what is important to you. That increased self-awareness will make job choices and job shifts easier. We pay a lot of attention to these issues in discussing the career paths in this book. The focus on the "fit" between you and your job may be especially helpful in determining the next steps in your career path.

Prepare yourself to be ready for job change. As college career counselors who see many adult alumni clients, we are all too familiar with the economic realities of the job search in today's job market. Not only has our society had to become accustomed to sudden bouts of layoffs, moves, and company closings because of mergers, changing markets, and general downsizing or outsourcing of jobs, but many workers have simply ignored the need to be prepared for such changes.

The most common laments we hear are "I haven't saved enough money to make a job change," and "I have so much debt I can't afford to take any less salary with a new job, no matter how great that new job is." These are sad comments on a situation that is easily remedied by reducing your consumption and increasing your savings to give you the needed flexibility to change jobs when you want to or have to. Take heed, and keep tight control over your personal finances. The best way to meet job-change challenges is to have zero debt and enough savings to cover your expenses for a while until you can get another good job.

Stay alert to the difference between contextual and portable skills. Contextual skills are those understandings, techniques, vocabularies, relationships, and factors that are very valuable and directly related to one job in a particular business but are of very little value outside your current field of work. For example, if you are working as an operations research analyst in the airline industry, your knowledge of aircraft loading weight will not be very helpful if you find your next job with a child's clothing manufacturer. In both jobs, however, your ability to analyze a delivery system and isolate potential problems and correct them does go with you from job to job.

The "portable skills" are those talents, knowledge, and techniques that you carry from job to job. You need both contextual and portable skills to succeed in any position, but you want to pay attention to the balance. If your employer offers training and that training is portable, take advantage of it.

All of our suggestions and advice are based on the precept that you enjoy and utilize your college degree to the maximum without sacrificing your own individuality or personal agenda. And it is possible! However, it doesn't just happen. It takes the kind of planning, researching, and exploring suggested by this book. More important, this process begins with you—knowing who you are and what you want at least for the next few years after graduation. Things will change as you gain experience in that first job after college and build a life for yourself. That's as it should be; but take the time right now to explore this guide, and use it to make your plan for the immediate future that you have been working and preparing for—it's almost here, at last!

PART ONE

THE JOB SEARCH

The Self-Assessment

Self-assessment is the process by which you begin to acknowledge your own particular blend of education, experiences, values, needs, and goals. It provides the foundation for career planning and the entire job search process. Self-assessment involves looking inward and asking yourself what can sometimes prove to be difficult questions. This self-examination should lead to an intimate understanding of your personal traits and values, consumption patterns and economic needs, longer-term goals, skill base, preferred skills, and underdeveloped skills.

You come to the self-assessment process knowing yourself well in some of these areas, but you may still be uncertain about other aspects. You may be well aware of your consumption patterns, but have you spent much time specifically identifying your longer-term goals or your personal values as they relate to work? No matter what level of self-assessment you have undertaken to date, it is now time to clarify all of these issues and questions as they relate to the job search.

The knowledge you gain in the self-assessment process will guide the rest of your job search. In this book, you will learn about all of the following tasks:

- Writing résumés and cover letters
- Researching careers and networking
- Interviewing and job offer considerations

In each of these steps, you will rely on and often return to the understanding gained through your self-assessment. Any individual seeking employment must be able and willing to express these facets of his or her personality to recruiters and interviewers throughout the job search. This communication allows you to show the world who you are so that together with employers you can determine whether there will be a workable match with a given job or career path.

How to Conduct a Self-Assessment

The self-assessment process goes on naturally all the time. People ask you to clarify what you mean, you make a purchasing decision, or you begin a new relationship. You react to the world and the world reacts to you. How you understand these interactions and any changes you might make because of them are part of the natural process of self-discovery. There is, however, a more comprehensive and efficient way to approach self-assessment with regard to employment.

Because self-assessment can become a complex exercise, we have distilled it into a seven-step process that provides an effective basis for undertaking a job search. The seven steps include the following:

1. Understanding your personal traits
2. Identifying your personal values
3. Calculating your economic needs
4. Exploring your longer-term goals
5. Enumerating your skill base
6. Recognizing your preferred skills
7. Assessing skills needing further development

As you work through your self-assessment, you might want to create a worksheet similar to the one shown in Exhibit 1.1, starting on the following page. Or you might want to keep a journal of the thoughts you have as you undergo this process. There will be many opportunities to revise your self-assessment as you start down the path of seeking a career.

Step 1   Understand Your Personal Traits
Each person has a unique personality that he or she brings to the job search process. Gaining a better understanding of your personal traits can help you evaluate job and career choices. Identifying these traits and then finding employment that allows you to draw on at least some of them can create a rewarding and fulfilling work experience. If potential employment doesn't allow you to use these preferred traits, it is important to decide whether you

Exhibit 1.1
SELF-ASSESSMENT WORKSHEET

Step 1. Understand Your Personal Traits

The personal traits that describe me are
(Include all of the words that describe you.)
The ten personal traits that most accurately describe me are
(List these ten traits.)

Step 2. Identify Your Personal Values

Working conditions that are important to me include
(List working conditions that would have to exist for you to accept a position.)
The values that go along with my working conditions are
(Write down the values that correspond to each working condition.)
Some additional values I've decided to include are
(List those values you identify as you conduct this job search.)

Step 3. Calculate Your Economic Needs

My estimated minimum annual salary requirement is
(Write the salary you have calculated based on your budget.)
Starting salaries for the positions I'm considering are
(List the name of each job you are considering and the associated starting salary.)

Step 4. Explore Your Longer-Term Goals

My thoughts on longer-term goals right now are
(Jot down some of your longer-term goals as you know them right now.)

Step 5. Enumerate Your Skill Base

The general skills I possess are
(List the skills that underlie tasks you are able to complete.)
The specific skills I possess are
(List more technical or specific skills that you possess, and indicate your level of expertise.)
General and specific skills that I want to promote to employers for the jobs I'm considering are
(List general and specific skills for each type of job you are considering.)

continued

Step 6. Recognize Your Preferred Skills

Skills that I would like to use on the job include

(List skills that you hope to use on the job, and indicate how often you'd like to use them.)

Step 7. Assess Skills Needing Further Development

Some skills that I'll need to acquire for the jobs I'm considering include

(Write down skills listed in job advertisements or job descriptions that you don't currently possess.)

I believe I can build these skills by

(Describe how you plan to acquire these skills.)

---

can find other ways to express them or whether you would be better off not considering this type of job. Interests and hobbies pursued outside of work hours can be one way to use personal traits you don't have an opportunity to draw on in your work. For example, if you consider yourself an outgoing person and the kinds of jobs you are examining allow little contact with other people, you may be able to achieve the level of interaction that is comfortable for you outside of your work setting. If such a compromise seems impractical or otherwise unsatisfactory, you probably should explore only jobs that provide the interaction you want and need on the job.

Many young adults who are not very confident about their employability will downplay their need for income. They will say, "Money is not all that important if I love my work." But if you begin to document exactly what you need for housing, transportation, insurance, clothing, food, and utilities, you will begin to understand that some jobs cannot meet your financial needs and it doesn't matter how wonderful the job is. If you have to worry each payday about bills and other financial obligations, you won't be very effective on the job. Begin now to be honest with yourself about your needs.

Begin the self-assessment process by creating an inventory of your personal traits. Make a list of as many words as possible to describe yourself. Words like *accurate, creative, future-oriented, relaxed,* or *structured* are just a few examples. In addition, you might ask people who know you well how they might describe you.

Focus on Selected Personal Traits. Of all the traits you identified, select the ten you believe most accurately describe you. Keep track of these ten traits.

Consider Your Personal Traits in the Job Search Process. As you begin exploring jobs and careers, watch for matches between your personal traits and the job descriptions you read. Some jobs will require many personal traits you know you possess, and others will not seem to match those traits.

Statisticians employ skills of data manipulation, analytical thinking, and research. It is important to realize that the field of statistics is essentially problem solving and not just number crunching. The ability to work on a team, analyze problems, and apply innovative solutions is a far more important personal trait for success than simply having a highly developed technical skill base. Statisticians must also possess self-discipline and time-management skills, since many projects are done independently. Statistics positions for new graduates often require the mastery of quantitative analysis. Attention to detail and excellent communication skills are also important attributes for this work.

Your ability to respond to changing conditions, your decision-making ability, productivity, creativity, and verbal skills all have a bearing on your success in and enjoyment of your work life. To better guarantee success, be sure to take the time needed to understand these traits in yourself.

Step 2   Identify Your Personal Values

Your personal values affect every aspect of your life, including employment, and they develop and change as you move through life. Values can be defined as principles that we hold in high regard, qualities that are important and desirable to us. Some values aren't ordinarily connected to work (love, beauty, color, light, relationships, family, or religion), and others are (autonomy, cooperation, effectiveness, achievement, knowledge, and security). Our values determine, in part, the level of satisfaction we feel in a particular job.

Define Acceptable Working Conditions. One facet of employment is the set of working conditions that must exist for someone to consider taking a job.

Each of us would probably create a unique list of acceptable working conditions, but items that might be included on many people's lists are the amount of money you would need to be paid, how far you are willing to drive or travel, the amount of freedom you want in determining your own schedule, whether you would be working with people or data or things, and the types of tasks you would be willing to do. Your conditions might include statements of working conditions you will *not* accept; for example, you might not be willing to work at night or on weekends or holidays.

If you were offered a job tomorrow, what conditions would have to exist for you to realistically consider accepting the position? Take some time and make a list of these conditions.

Realize Associated Values. Your list of working conditions can be used to create an inventory of your values relating to jobs and careers you are exploring. For example, if one of your conditions stated that you wanted to earn at least $30,000 per year, the associated value would be financial gain. If another condition was that you wanted to work with a friendly group of people, the value that went along with that might be belonging or interaction with people.

Relate Your Values to the World of Work. As you read the job descriptions you come across either in this book, in newspapers and magazines, or online, think about the values associated with each position.

As a statistician, your duties might include projects such as designing studies and collecting, analyzing, and interpreting data for the planning of service provisions in a government agency; analyzing data in the investigation of effects of a new drug for AIDS; or preparing a statistical analysis of risks in development of a new product line and its projected profits or losses for the first, second, and third years of its production.

At least some of the associated values in the field you're exploring should match those you extracted from your list of working conditions. Take a second look at any values that don't match up. How important are they to you? What will happen if they are not satisfied on the job? Can you incorporate those personal values elsewhere? Your answers need to be brutally honest. As you continue your exploration, be sure to add to your list any additional values that occur to you.

Step 3   Calculate Your Economic Needs

Each of us grew up in an environment that provided for certain basic needs, such as food and shelter, and, to varying degrees, other needs that we now consider basic, such as cable television, e-mail, or an automobile. Needs such as privacy, space, and quiet, which at first glance may not appear to be monetary needs, may add to housing expenses and so should be considered as you examine your economic needs. For example, if you place a high value on a large, open living space for yourself, it would be difficult to satisfy that need without an associated high housing cost, especially in a densely populated city environment.

As you prepare to move into the world of work and become responsible for meeting your own basic needs, it is important to consider the salary you will need to be able to afford a satisfying standard of living. The three-step process outlined here will help you plan a budget, which in turn will allow you to evaluate the various career choices and geographic locations you are considering. The steps include (1) develop a realistic budget, (2) examine starting salaries, and (3) use a cost-of-living index.

Develop a Realistic Budget. Each of us has certain expectations for the kind of lifestyle we want to maintain. To begin the process of defining your economic needs, it will be helpful to determine what you expect to spend on routine monthly expenses. These expenses include housing, food, transportation, entertainment, utilities, loan repayments, and revolving charge accounts. You may not currently spend anything for certain items, but you probably will have to once you begin supporting yourself. As you develop this budget, be generous in your estimates, but keep in mind any items that could be reduced or eliminated. If you are not sure about the cost of a certain item, talk with family or friends who would be able to give you a realistic estimate.

If this is new or difficult for you, start to keep a log of expenses right now. You may be surprised at how much you actually spend each month for food or stamps or magazines. Household expenses and personal grooming items can often loom very large in a budget, as can auto repairs or home maintenance.

Income taxes must also be taken into consideration when examining salary requirements. State and local taxes vary, so it is difficult to calculate exactly the effect of taxes on the amount of income you need to generate. To roughly estimate the gross income necessary to generate your minimum annual salary requirement, multiply the minimum salary you have calculated by a factor of 1.35. The resulting figure will be an approximation of what your gross income would need to be, given your estimated expenses.

Examine Starting Salaries. Starting salaries for each of the career tracks are provided throughout this book. These salary figures can be used in conjunction with the cost-of-living index (discussed in the next section) to determine whether you would be able to meet your basic economic needs in a given geographic location.

Use a Cost-of-Living Index. If you are thinking about trying to get a job in a geographic region other than the one where you now live, understanding differences in the cost of living will help you come to a more informed decision about making a move. By using a cost-of-living index, you can compare salaries offered and the cost of living in different locations with what you know about the salaries offered and the cost of living in your present location.

Many variables are used to calculate the cost-of-living index. Often included are housing, groceries, utilities, transportation, health care, clothing, and entertainment expenses. Right now you do not need to worry about the details associated with calculating a given index. The main purpose of this exercise is to help you understand that pay ranges for entry-level positions may not vary greatly, but the cost of living in different locations *can* vary tremendously.

For example, if you lived in Des Moines, Iowa, and you were interested in working as an actuary in the insurance industry, you might earn, at entry level, about $40,000 annually. But let's say you're also thinking about moving to Boston, Denver, or Omaha. You know that you can live well enough on $40,000 in Des Moines, but you want to be able to equal that lifestyle with your salary in the other locations you're considering. How much will you need to earn in those locations to do this? Checking the cost-of-living index for each city will show you.

Let's walk through an example. In any cost-of-living index, the number 100 represents the national average cost of living, and each city is assigned an index number based on the current prices in that city for the items included in the index, such as housing, food, and so on. In the 2005 index that we used, Boston was assigned the number 135.5, Denver's index was 102.9, Omaha's was 89.2, and Des Moines's index was 91.4.

In other words, it costs almost 50 percent more to live in Boston than it does in Des Moines, and Denver is more expensive as well. We can set up a table to determine exactly how much you would have to earn in each of these cities to have the same buying power that you have in Des Moines.

JOB: ACTUARY (ENTRY LEVEL)

City	Index	Equivalent Salary
Boston	135.5	
		$\dfrac{135.5}{91.4} \times \$40,000 = \$59,200$ in Boston
Des Moines	91.4	
Denver	102.9	
		$\dfrac{102.9}{91.4} \times \$40,000 = \$45,200$ in Denver
Des Moines	91.4	
Omaha	89.2	
		$\dfrac{89.2}{91.4} \times \$40,000 = \$38,200$ in Omaha
Des Moines	91.4	

You would have to earn $59,200 in Boston, $45,200 in Denver, and $38,200 in Omaha to match the buying power of $40,000 in Des Moines.

If you would like to determine whether it is financially worthwhile to make any of these moves, one more piece of information is needed—the salaries of the actuaries in these other cities. Internet listings for job openings in 2005 in these cities showed salary ranges for "entry-level" or "entry-level actuarial student" as the following: Boston, from $50,000 to $75,000; Denver, from $45,000 to $60,000; and Omaha, from $35,000 to $50,000.

Based on these figures, you would have to negotiate a higher salary in Boston or Denver in order to match your buying power in Des Moines. In each of these cities, the current salary ranges being offered would allow you to do that, but it would be up to you to make choices and negotiate successfully in order to make such a move worthwhile.

You can work through a similar exercise for any type of job you are considering and for many locations when current salary information is available. It will be worth your time to undertake this analysis if you are seriously considering a relocation. By doing so you will be able to make an informed choice.

Step 4   Explore Your Longer-Term Goals

There is no question that when we first begin working, our goals are to use our skills and education in a job that will reward us with employment, income, and status relative to the preparation we brought with us to this position. If we are not being paid as much as we feel we should for our level of education or if job demands don't provide the intellectual stimulation we had hoped for, we experience unhappiness and as a result often seek other employment.

Most jobs we consider "good" are those that fulfill our basic "lower-level" needs of security, food, clothing, shelter, income, and productive work. But even when our basic needs are met and our jobs are secure and productive, we as individuals are constantly changing. As we change, the demands and expectations we place on our jobs may change. Fortunately, some jobs grow and change with us, and this explains why some people are happy throughout many years in a job.

But more often people are bigger than the jobs they fill. We have more goals and needs than any job could satisfy. These are "higher-level" needs of self-esteem, companionship, affection, and an increasing desire to feel we are employing ourselves in the most effective way possible. Not all of these higher-level needs can be met through employment, but for as long as we are employed, we increasingly demand that our jobs play their part in moving us along the path to fulfillment.

Another obvious but important fact is that we change as we mature. Although our jobs also have the potential for change, they may not change as frequently or as markedly as we do. There are increasingly fewer one-job, one-employer careers; we must think about a work future that may involve voluntary or forced moves from employer to employer. Because of that very real possibility, we need to take advantage of the opportunities in each position we hold. Acquiring the skills and competencies associated with each position will keep us viable and attractive as employees. This is particularly true in a job market that not only is technology/computer dependent, but also is populated with more and more small, self-transforming organizations rather than the large, seemingly stable organizations of the past.

Talking with experienced people in your intended field can be a rich source of insight. If you are considering a position as a retail buyer, you will gain a better perspective on this career if you can

talk with an entry-level associate buyer, a more senior and experienced department head, and then finally, a vice president of
store operations or sales merchandising who has had a considerable work history in the retail sector. Each one will have a different perspective, unique concerns, and an individual set of
value priorities.

Step 5 Enumerate Your Skill Base

In terms of the job search, skills can be thought of as capabilities that can
be developed in school, at work, or by volunteering and then used in specific job settings. Many studies have documented the kinds of skills that
employers seek in entry-level applicants. For example, some of the most
desired skills for individuals interested in the teaching profession are the ability to interact effectively with students one-on-one, to manage a classroom,
to adapt to varying situations as necessary, and to get involved in school activities. Business employers have also identified important qualities, including
enthusiasm for the employer's product or service, a businesslike mind, the
ability to follow written or oral instructions, the ability to demonstrate self-
control, the confidence to suggest new ideas, the ability to communicate with
all members of a group, an awareness of cultural differences, and loyalty, to
name just a few. You will find that many of these skills are also in the repertoire of qualities demanded in your college major.

To be successful in obtaining any given job, you must be able to demonstrate that you possess a certain mix of skills that will allow you to carry out
the duties required by that job. This skill mix will vary a great deal from job
to job; to determine the skills necessary for the jobs you are seeking, you
can read job advertisements or more generic job descriptions, such as those
found later in this book. If you want to be effective in the job search, you
must directly show employers that you possess the skills needed to be successful in filling the position. These skills will initially be described on your
résumé and then discussed again during the interview process.

Skills are either general or specific. To develop a list of skills relevant to
employers, you must first identify the general skills you possess, then list
specific skills you have to offer, and, finally, examine which of these skills
employers are seeking.

Identify Your General Skills. Because you possess or will possess a college
degree, employers will assume that you can read and write, perform certain
basic computations, think critically, and communicate effectively. Employ-

ers will want to see that you have acquired these skills, and they will want to know which additional general skills you possess.

One way to begin identifying skills is to write an experiential diary. An experiential diary lists all the tasks you were responsible for completing for each job you've held and then outlines the skills required to do those tasks. You may list several skills for any given task. This diary allows you to distinguish between the tasks you performed and the underlying skills required to complete those tasks. Here's an example:

Tasks	Skills
Answering telephone	Effective use of language, clear diction, ability to direct inquiries, ability to solve problems
Waiting on tables	Poise under conditions of time and pressure, speed, accuracy, good memory, simultaneous completion of tasks, sales skills

For each job or experience you have participated in, develop a worksheet based on the example shown here. On a résumé, you may want to describe these skills rather than simply listing tasks. Skills are easier for the employer to appreciate, especially when your experience is very different from the employment you are seeking. In addition to helping you identify general skills, this experiential diary will prepare you to speak more effectively in an interview about the qualifications you possess.

Identify Your Specific Skills. It may be easier to identify your specific skills because you can definitely say whether you can speak other languages, program a computer, draft a map or diagram, or edit a document using appropriate symbols and terminology.

Using your experiential diary, identify the points in your history where you learned how to do something very specific, and decide whether you have a beginning, intermediate, or advanced knowledge of how to use that particular skill. Right now, be sure to list *every* specific skill you have, and don't consider whether you like using the skill. Write down a list of specific skills you have acquired and the level of competence you possess—beginning, intermediate, or advanced.

Relate Your Skills to Employers. You probably have thought about a couple of different jobs you might be interested in obtaining, and one way to begin relating the general and specific skills you possess to a potential employer's needs is to read actual advertisements for these types of positions (see Part Two for resources listing actual job openings).

For example, you might be interested in working as an actuary within a large insurance company, prior to returning to graduate school for your master of science (MS) degree. A typical job listing might read, "Actuary—Insurance: Applicant should have expertise performing statistical and actuarial analysis on personal line insurance. Completion of one or more actuarial exams preferred. The applicant should be comfortable with Windows, Excel, and Word. Basic knowledge of insurance accounting concepts required. This position requires a four-year degree in actuarial science, computer science, accounting, mathematics, or equivalent, along with superior knowledge of statistical methods. The work will involve embedded value analysis, cash flow testing, budgeting, and pricing, with other assistance related to financial reporting required on a seasonal basis."

A sample list of skills that are commonly needed for actuarial work in the insurance industry is shown below. This list has been derived from actual job listings and industry job descriptions.

JOB: ACTUARY—INSURANCE COMPANY

General Skills	Specific Skills
Computer skills	Analyze insurance rates and loss
Gathering information	reserves
Decision making	Test system changes
Meeting deadlines	Prepare rate and statistical filings for
Reading	submission to regulatory agencies
Writing	Develop market share and CPI studies
Collaborating on projects	Evaluate alternatives
Attending meetings	Master various software packages
	Mathematical computations
	Generate ratios

Using separate sheets of paper, generate both a general skills list and a specific skills list for at least one job that you are considering. Use any job advertisements, books, Internet, and other sources to help you generate your lists. Most of the items on your general skills list are very portable skills—you can transfer them to many jobs. You will probably find that many of the skills on your specific skills list are also portable. For example, evaluating alternatives is a required skill for actuaries, mathematicians, and operations researchers, and would be helpful for statisticians, as well.

Step 6 Recognize Your Preferred Skills

In the previous section you developed a comprehensive list of skills that relate to particular career paths that are of interest to you. You can now relate these to skills that you prefer to use. We all use a wide range of skills (some researchers say individuals have a repertoire of about five hundred skills), but we may not particularly be interested in using all of them in our work. There may be some skills that come to us more naturally or that we use successfully time and time again and that we want to continue to use; these are best described as our preferred skills. For this exercise use the list of skills that you created for the previous section, and decide which of them you are *most interested in using* in future work and how often you would like to use them. You might be interested in using some skills only occasionally, while others you would like to use more regularly. You probably also have skills that you hope you can use constantly.

As you examine job announcements, look for matches between this list of preferred skills and the qualifications described in the advertisements. These skills should be highlighted on your résumé and discussed in job interviews.

Step 7 Assess Skills Needing Further Development

Previously you compiled a list of general and specific skills required for given positions. You already possess some of these skills; those that remain to be developed are your underdeveloped skills.

If you are just beginning the job search, there may be gaps between the qualifications required for some of the jobs you're considering and the skills you possess. The thought of having to admit to and talk about these underdeveloped skills, especially in a job interview, is a frightening one. One way to put a healthy perspective on this subject is to target and relate your exploration of underdeveloped skills to the types of positions you are seeking. Recognizing these shortcomings and planning to overcome them with either on-the-job training or additional formal education can be a positive way to address the concept of underdeveloped skills.

On your worksheet or in your journal, make a list of up to five general or specific skills required for the positions you're interested in that you *don't currently possess*. For each item list an idea you have for specific action you could take to acquire that skill. Do some brainstorming to come up with possible actions. If you have a hard time generating ideas, talk to people currently working in this type of position, professionals in your college career services office, trusted friends, family members, or members of related professional associations.

In the chapter on interviewing, we will discuss in detail how to effectively address questions about underdeveloped skills. Generally speaking, though, employers want genuine answers to these types of questions. They want you to reveal "the real you," and they also want to see how you answer difficult questions. In taking the positive, targeted approach discussed previously, you show the employer that you are willing to continue to learn and that you have a plan for strengthening your job qualifications.

Use Your Self-Assessment

Exploring entry-level career options can be an exciting experience if you have good resources available and will take the time to use them. Can you effectively complete the following tasks?

1. Understand your personality traits and relate them to career choices
2. Define your personal values
3. Determine your economic needs
4. Explore longer-term goals
5. Understand your skill base
6. Recognize your preferred skills
7. Express a willingness to improve on your underdeveloped skills

If so, then you can more meaningfully participate in the job search process by writing a more effective résumé, finding job titles that represent work you are interested in doing, locating job sites that will provide the opportunity for you to use your strengths and skills, networking in an informed way, participating in focused interviews, getting the most out of follow-up contacts, and evaluating job offers to find those that create a good match between you and the employer. The remaining chapters in Part One guide you through these next steps in the job search process. For many job seekers, this process can take anywhere from three months to a year to implement. The time you will need to put into your job search will depend on the type of job you want and the geographic location where you'd like to work. Think of your effort as a job in itself, requiring you to set aside time each week to complete the needed work. Carefully undertaken efforts may reduce the time you need for your job search.

The Résumé and Cover Letter

The task of writing a résumé may seem overwhelming if you are unfamiliar with this type of document, but there are some easily understood techniques that can and should be used. This section was written to help you understand the purpose of the résumé, the different types of formats available, and how to write the sections that contain information traditionally found on a résumé. We will present examples and explanations that address questions frequently posed by people writing their first résumé or updating an old one.

Even within the formats and suggestions given, however, there are infinite variations. True, most follow one of the outlines suggested, but you should feel free to adjust the résumé to suit your needs and make it expressive of your life and experience.

Why Write a Résumé?

The purpose of a résumé is to convince an employer that you should be interviewed. Whether you're mailing, faxing, or e-mailing this document, you'll want to present enough information to show that you can make an immediate and valuable contribution to an organization. A résumé is not an in-depth historical or legal document; later in the job search process you may be asked to document your entire work history on an application form and attest to its validity. The résumé should, instead, highlight relevant information pertaining directly to the organization that will receive the document or to the type of position you are seeking.

We will discuss the chronological and digital résumés in detail here. Functional and targeted résumés, which are used much less often, are briefly discussed. The reasons for using one type of résumé over another and the typical format for each are addressed in the following sections.

The Chronological Résumé

The chronological résumé is the most common of the various résumé formats and therefore the format that employers are most used to receiving. This type of résumé is easy to read and understand because it details the chronological progression of jobs you have held. (See Exhibit 2.1.) It begins with your most recent employment and works back in time. If you have a solid work history or have experience that provided growth and development in your duties and responsibilities, a chronological résumé will highlight these achievements. The typical elements of a chronological résumé include the heading, a career objective, educational background, employment experience, activities, and references.

The Heading
The heading consists of your name, address, telephone number, and other means of contact. This may include a fax number, e-mail address, and your home-page address. If you are using a shared e-mail account or a parent's business fax, be sure to let others who use these systems know that you may receive important professional correspondence via these systems. You wouldn't want to miss a vital e-mail or fax! Likewise, if your résumé directs readers to a personal home page on the Web, be certain it's a professional personal home page designed to be viewed and appreciated by a prospective employer. This may mean making substantial changes in the home page you currently mount on the Web.

The Objective
Without a doubt the objective statement is the most challenging part of the résumé for most writers. Even for individuals who have decided on a career path, it can be difficult to encapsulate all they want to say in one or two brief sentences. For job seekers who are unfocused or unclear about their intentions, trying to write this section can inhibit the entire résumé writing process.

Keep the objective as short as possible and no longer than two short sentences.

Exhibit 2.1
CHRONOLOGICAL RÉSUMÉ

MARTINE VENDEZ

Bolder Hall #306   43 River Road
Selbert State College   Benton, New Hampshire 04523
Selberton, Vermont 01234   (603) 555-1111
(802) 555-4567   (after May 28, 2006)

OBJECTIVE

A high school mathematics teaching position where I can motivate the students to meet the challenges and overcome any difficulties they might meet in all levels of mathematics.

EDUCATION

Bachelor of Science Degree in Mathematics Education, Selbert State College, Selberton, Vermont, May 2006

HONORS

President's List, Dean's List, Kappa Delta Phi Education Honor Society, Selbert State Scholar, Outstanding Student of Education

PROFESSIONAL EXPERIENCE

Student Teacher, Selbert Institute, January–May 2006
Assisted in teaching mathematics in classes ranging in size from fifteen to thirty students, grades nine through twelve. Administrative duties included constructing lesson plans, recording grades, helping prepare and send progress reports, and maintaining daily attendance records. Instructed daily lessons in Algebra, Geometry, and Advanced Math (Pre-Calculus). Attended all faculty and department meetings and workshops, and served as junior class adviser.

Help Session Instructor, Selbert State College, 2003–05
Was employed by Mathematics Department to assist college students who had difficulty in Fundamental Mathematics (Pre-Algebra) and Algebra. Held group sessions to aid students in preparing for exams.

continued

Mathematics Tutor, Selbert State College, 2001–03
Tutored individuals and groups of students in various mathematics courses such as Fundamental Math (Pre-Algebra), Algebra, Statistics I, Discrete Math, Elementary Functions (Pre-Calculus), Calculus, and Math Lab Activities.

Part-Time Sales Associate, Best Foods, Glendale, Vermont, 2001–04
Organized stock in produce department and submitted accurate, daily records in a timely manner. Interacted daily with customers and employees. Was twice selected as Sales Associate of the Month.

PROFESSIONAL MEMBERSHIPS
National Council of Teachers of Mathematics, Future Educators of America

Choose one of the following types of objective statement:

1. *General Objective Statement*

- An entry-level educational programming coordinator position

2. *Position-Focused Objective*

- To obtain the position of conference coordinator at State College

3. *Industry-Focused Objective*

- To begin a career as a sales representative in the cruise line industry

4. *Summary of Qualifications Statement*

A bachelor of science degree in mathematics and four years of progressively increasing job responsibility as a statistical analyst have prepared me to begin a career as a mathematical statistician for a company that values a consistently high level of dedication, and strong analytical and design skills.

Support Your Objective. A résumé that contains any one of these types of objective statements should then go on to demonstrate why you are qualified to get the position. Listing academic degrees can be one way to indicate qualifications. Another demonstration would be in the way previous experiences, both volunteer and paid, are described. Without this kind of documentation in the body of the résumé, the objective looks unsupported. Think of the résumé as telling a connected story about you. All the elements should work together to form a coherent picture that ideally should relate to your statement of objective.

Education

This section of your résumé should indicate the exact name of the degree you will receive or have received, spelled out completely with no abbreviations. The degree is generally listed after the objective, followed by the institution name and location, and then the month and year of graduation. This section could also include your academic minor, grade point average (GPA), and appearance on the Dean's List or President's List.

If you have enough space, you might want to include a section listing courses related to the field in which you are seeking work. The best use of a "related courses" section would be to list some course work that is not traditionally associated with the major. Perhaps you took several computer courses outside your degree that will be helpful and related to the job prospects you are entertaining. Several education section examples are shown here:

- Bachelor of Science Degree in Mathematics; State University, Des Moines, Iowa; May 2005; Concentration: Actuarial Math
- Bachelor of Science Degree in Mathematics Education; State College, Keene, New Hampshire; May 2005; Option: Middle/Junior High School
- Bachelor of Science Degree in Mathematics; Community College, Baltimore, Maryland; May 2005; Minor: Operations Research

An example of a format for a related courses section follows:

RELATED COURSES

Statistics I and II	Numerical Methods
Calculus I and II	Using the Computer
Elements of Linear Algebra with Discrete Mathematics	Quantitative Methods Business Applications
Probability Theory	Time and Money

Experience

The experience section of your résumé should be the most substantial part and should take up most of the space on the page. Employers want to see what kind of work history you have. They will look at your range of experiences, longevity in jobs, and specific tasks you are able to complete. This section may also be called "work experience," "related experience," "employment history," or "employment." No matter what you call this section, some important points to remember are the following:

1. **Describe your duties** as they relate to the position you are seeking.
2. **Emphasize major responsibilities** and indicate increases in responsibility. Include all relevant employment experiences: summer, part-time, internships, cooperative education, or self-employment.
3. **Emphasize skills**, especially those that transfer from one situation to another. The fact that you coordinated a student organization, chaired meetings, supervised others, and managed a budget leads one to suspect that you could coordinate other things as well.
4. **Use descriptive job titles** that provide information about what you did. A "Student Intern" should be more specifically stated as, for example, "Magazine Operations Intern." "Volunteer" is also too general; a title such as "Peer Writing Tutor" would be more appropriate.
5. **Create word pictures** by using active verbs to start sentences. Describe *results* you have produced in the work you have done.

A limp description would say something such as the following: "My duties included helping with production, proofreading, and editing. I used a design and page layout program." An action statement would be stated as follows: "Coordinated and assisted in the creative marketing of brochures and seminar promotions, becoming proficient in Quark."

Remember, an accomplishment is simply a result, a final measurable product that people can relate to. A duty is not a result; it is an obligation—every job holder has duties. For an effective résumé, list as many results as you can. To make the most of the limited space you have and to give your description impact, carefully select appropriate and accurate descriptors.

Here are some traits that employers tell us they like to see:

- Teamwork
- Energy and motivation

- Learning and using new skills
- Versatility
- Critical thinking
- Understanding how profits are created
- Organizational acumen
- Communicating directly and clearly, in both writing and speaking
- Risk taking
- Willingness to admit mistakes
- High personal standards

Solutions to Frequently Encountered Problems

Repetitive Employment with the Same Employer
EMPLOYMENT: The Foot Locker, Portland, Oregon. Summer 2001, 2002, 2003. Initially employed in high school as salesclerk. Because of successful performance, asked to return next two summers at higher pay with added responsibility. Ranked as the #2 salesperson the first summer and #1 the next two summers. Assisted in arranging eye-catching retail displays; served as manager of other summer workers during owner's absence.

A Large Number of Jobs
EMPLOYMENT: Recent Hospitality Industry Experience: Affiliated with four upscale hotel/restaurant complexes (September 2001–February 2004), where I worked part- and full-time as a waiter, bartender, disc jockey, and bookkeeper to produce income for college.

Several Positions with the Same Employer
EMPLOYMENT: Coca-Cola Bottling Co., Burlington, Vermont, 2001–2004. In four years, I received three promotions, each with increased pay and responsibility.

Summer Sales Coordinator: Promoted to hire, train, and direct efforts of add-on staff of fifteen college-age route salespeople hired to meet summer peak demand for product.

Sales Administrator: Promoted to run home office sales desk, managing accounts and associated delivery schedules for professional sales force of ten people. Intensive phone work, daily interaction with all personnel, and strong knowledge of product line required.

Route Salesperson: Summer employment to travel and tourism industry sites that use Coke products. Met specific schedule demands, used good communication skills with wide variety of customers, and demonstrated strong selling skills. Named salesperson of the month for July and August of that year.

Questions Résumé Writers Often Ask

How Far Back Should I Go in Terms of Listing Past Jobs?
Usually, listing three or four jobs should suffice. If you did something back in high school that has a bearing on your future aspirations for employment, by all means list the job. As you progress through your college career, high school jobs will be replaced on the résumé by college employment.

Should I Differentiate Between Paid and Nonpaid Employment?
Most employers are not initially concerned about how much you were paid. They are eager to know how much responsibility you held in your past employment. There is no need to specify that your work was as a volunteer if you had significant responsibilities.

How Should I Represent My Accomplishments or Work-Related Responsibilities?
Succinctly, but fully. In other words, give the employer enough information to arouse curiosity but not so much detail that you leave nothing to the imagination. Besides, some jobs merit more lengthy explanations than others. Be sure to convey any information that can give an employer a better understanding of the depth of your involvement at work. Did you supervise others? How many? Did your efforts result in a more efficient operation? How much did you increase efficiency? Did you handle a budget? How much? Were you promoted in a short time? Did you work two jobs at once or fifteen hours per week after high school? Where appropriate, quantify.

Should the Work Section Always Follow the Education Section on the Résumé?
Always lead with your strengths. If your education closely relates to the employment you now seek, put this section after the objective. If your edu-

cation does not closely relate but you have a surplus of good work experiences, consider reversing the order of your sections to lead with employment, followed by education.

How Should I Present My Activities, Honors, Awards, Professional Societies, and Affiliations?

This section of the résumé can add valuable information for an employer to consider if used correctly. The rule of thumb for information in this section is to include only those activities that are in some way relevant to the objective stated on your résumé. If you can draw a valid connection between your activities and your objective, include them; if not, leave them out.

Professional affiliations and honors should all be listed; especially important are those related to your job objective. Social clubs and activities need not be a part of your résumé unless you hold a significant office or you are looking for a position related to your membership. Be aware that most prospective employers' principal concerns are related to your employability, not your social life. If you have any, publications can be included as an addendum to your résumé.

How Should I Handle References?

The use of references is considered a part of the interview process, and they should never be listed on a résumé. You would always provide references to a potential employer if requested to, so it is not even necessary to include this section on the résumé if space does not permit. If space is available, it is acceptable to include the following statement:

- References furnished upon request.

The Functional Résumé

A functional résumé departs from a chronological résumé in that it organizes information by specific accomplishments in various settings: previous jobs, volunteer work, associations, and so forth. This type of résumé permits you to stress the substance of your experiences rather than the position titles you have held. You should consider using a functional résumé if you have held a series of similar jobs that relied on the same skills or abilities. There are many good books in which you can find examples of functional résumés, including *How to Write a Winning Resume* or *Resumes Made Easy*.

The Targeted Résumé

The targeted résumé focuses on specific work-related capabilities you can bring to a given position within an organization. Past achievements are listed to highlight your capabilities and the work history section is abbreviated.

Digital Résumés

Today's employers have to manage an enormous number of résumés. One of the most frequent complaints the writers of this series hear from students is the failure of employers to even acknowledge the receipt of a résumé and cover letter. Frequently, the reason for this poor response or nonresponse is the volume of applications received for every job. In an attempt to better manage the considerable labor investment involved in processing large numbers of résumés, many employers are requiring digital submission of résumés. There are two types of digital résumés: those that can be e-mailed or posted to a website, called *electronic résumés*, and those that can be "read" by a computer, commonly called *scannable résumés*. Though the format may be a bit different from the traditional "paper" résumé, the goal of both types of digital résumés is the same—to get you an interview! These résumés must be designed to be "technologically friendly." What that basically means to you is that they should be free of graphics and fancy formatting. (See Exhibit 2.2.)

Electronic Résumés

Sometimes referred to as plain-text résumés, electronic résumés are designed to be e-mailed to an employer or posted to one of many commercial Internet databases such as CareerMosaic.com, America's Job Bank (ajb.dni.us), or Monster.com.

Some technical considerations:

- Electronic résumés must be written in American Standard Code for Information Interchange (ASCII), which is simply a plain-text format. These characters are universally recognized so that every computer can accurately read and understand them. To create an ASCII file of your current résumé, open your document, then save it as a text or ASCII file. This will eliminate all formatting. Edit as needed using your computer's text editor application.

Exhibit 2.2
DIGITAL RÉSUMÉ

MARTINE VENDEZ ◄───────────────────────── Put your name at the
Selbert State College   top on its own line.
Selberton, Vermont 01234
(802) 555-4567 ◄───────────────────────────
mvendez@xxx.com ─────────────────────────── Put your phone number
                                             on its own line.

KEYWORD SUMMARY
Mathematics Teacher
Secondary Education   ◄──────── Keywords make your
Algebra and Geometry   résumé easier to find in
Pre-Calculus   a database.

EXPERIENCE ───────────────────────────── Use a space between
* Student Teaching January-May 2006;   asterisk and text.
Selbert Institute.
Teach mathematics in grades nine through
twelve. Construct lesson plans, grade papers, No line should exceed
prepare progress reports. Instruct lessons   sixty-five characters.
in Algebra, Geometry, Pre-Calculus.
                                              End each line by
                                              hitting the ENTER
                                              (or RETURN) key.

CERTIFICATION ◄─────────────────────────
State of Vermont Grades 5-12   Capitalize letters to
Mathematics Education   emphasize headings.

- Use a standard-width typeface. Courier is a good choice because it is the font associated with ASCII in most systems.
- Use a font size of 11 to 14 points. A 12-point font is considered standard.
- Your margin should be left-justified.
- Do not exceed sixty-five characters per line because the word-wrap function doesn't operate in ASCII.
- Do not use boldface, italics, underlining, bullets, or various font sizes. Instead, use asterisks, plus signs, or all capital letters when you want to emphasize something.
- Avoid graphics and shading.
- Use as many "keywords" as you possibly can. These are words or phrases usually relating to skills or experience that either are specifically used in the job announcement or are popular buzzwords in the industry.
- Minimize abbreviations.
- Your name should be the first line of text.
- Conduct a "test run" by e-mailing your résumé to yourself and a friend before you send it to the employer. See how it transmits, and make any changes you need to. Continue to test it until it's exactly how you want it to look.
- Unless an employer specifically requests that you send the résumé in the form of an attachment, don't. Employers can encounter problems opening a document as an attachment, and there are always viruses to consider.
- Don't forget your cover letter. Send it along with your résumé as a single message.

Scannable Résumés

Some companies are relying on technology to narrow the candidate pool for available job openings. Electronic Applicant Tracking uses imaging to scan, sort, and store résumé elements in a database. Then, through OCR (Optical Character Recognition) software, the computer scans the résumés for keywords and phrases. To have the best chance at getting an interview, you want to increase the number of "hits"—matches of your skills, abilities, experience, and education to those the computer is scanning for—your résumé will get. You can see how critical using the right keywords is for this type of résumé.

Technical considerations include:

- Again, do not use boldface (newer systems may be able to read this, but many older ones won't), italics, underlining, bullets, shading, graphics, or multiple font sizes. Instead, for emphasis, use asterisks, plus signs, or all capital letters. Minimize abbreviations.
- Use a popular typeface such as Courier, Helvetica, Ariel, or Palatino. Avoid decorative fonts.
- Font size should be between 11 and 14 points.
- Do not compress the spacing between letters.
- Use horizontal and vertical lines sparingly; the computer may misread them as the letters *L* or *I*.
- Left-justify the text.
- Do not use parentheses or brackets around telephone numbers, and be sure your phone number is on its own line of text.
- Your name should be the first line of text and on its own line. If your résumé is longer than one page, be sure to put your name on the top of all pages.
- Use a traditional résumé structure. The chronological format may work best.
- Use nouns that are skill-focused, such as *management, writer,* and *programming.* This is different from traditional paper résumés, which use action-oriented verbs.
- Laser printers produce the finest copies. Avoid dot-matrix printers.
- Use standard, light-colored paper with text on one side only. Since the higher the contrast, the better, your best choice is black ink on white paper.
- Always send original copies. If you must fax, set the fax on fine mode, not standard.
- Do not staple or fold your résumé. This can confuse the computer.
- Before you send your scannable résumé, be certain the employer uses this technology. If you can't determine this, you may want to send two versions (scannable and traditional) to be sure your résumé gets considered.

Résumé Production and Other Tips

An ink-jet printer is the preferred option for printing your résumé. Begin by printing just a few copies. You may find a small error or you may simply want to make some changes, and it is less frustrating and less expensive if you print in small batches.

Résumé paper color should be carefully chosen. You should consider the types of employers who will receive your résumé and the types of positions for which you are applying. Use white or ivory paper for traditional or conservative employers or for higher-level positions.

Black ink on sharp, white paper can be harsh on the reader's eyes. Think about an ivory or cream paper that will provide less contrast and be easier to read. Pink, green, and blue tints should generally be avoided.

Many résumé writers buy packages of matching envelopes and cover sheet stationery that, although not absolutely necessary, help convey a professional impression.

If you'll be producing many cover letters at home, be sure you have high-quality printing equipment. Learn standard envelope formats for business, and retain a copy of every cover letter you send out. You can use the copies to take notes of any telephone conversations that may occur.

If attending a job fair, either carry a briefcase or place your résumé in a nicely covered legal-size pad holder.

The Cover Letter

The cover letter provides you with the opportunity to tailor your résumé by telling the prospective employer how you can be a benefit to the organization. It allows you to highlight aspects of your background that are not already discussed in your résumé and that might be especially relevant to the organization you are contacting or to the position you are seeking. Every résumé should have a cover letter enclosed when you send it out. Unlike the résumé, which may be mass-produced, a cover letter is most effective when it is individually prepared and focused on the particular requirements of the organization in question.

A good cover letter should supplement the résumé and motivate the reader to review the résumé. The format shown in Exhibit 2.3 (see page 34) is only a suggestion to help you decide what information to include in a cover letter.

Begin the cover letter with your street address six lines down from the top. Leave three to five lines between the date and the name of the person to whom you are addressing the cover letter. Make sure you leave one blank line between the salutation and the body of the letter and between paragraphs. After typing "Sincerely," leave four blank lines and type your name.

This should leave plenty of room for your signature. A sample cover letter is shown in Exhibit 2.4 on page 35.

The following guidelines will help you write good cover letters:

1. Be sure to type your letter neatly; ensure there are no misspellings.
2. Avoid unusual typefaces, such as script.
3. Address the letter to an individual, using the person's name and title. To obtain this information, call the company. If answering a blind newspaper advertisement, address the letter "To Whom It May Concern" or omit the salutation.
4. Be sure your cover letter directly indicates the position you are applying for and tells why you are qualified to fill it.
5. Send the original letter, not a photocopy, with your résumé. Keep a copy for your records.
6. Make your cover letter no more than one page.
7. Include a phone number where you can be reached.
8. Avoid trite language and have someone read the letter over to react to its tone, content, and mechanics.
9. For your own information, record the date you send out each letter and résumé.

Exhibit 2.3
COVER LETTER FORMAT

Your Street Address
Your Town, State, Zip
Phone Number
Fax Number
E-mail

Date

Name
Title
Organization
Address

Dear _____:

First Paragraph. In this paragraph state the reason for the letter, name the specific position or type of work you are applying for, and indicate from which resource (career services office, website, newspaper, contact, employment service) you learned of this opening. The first paragraph can also be used to inquire about future openings.

Second Paragraph. Indicate why you are interested in this position, the company, or its products or services and what you can do for the employer. If you are a recent graduate, explain how your academic background makes you a qualified candidate. Try not to repeat the same information found in the résumé.

Third Paragraph. Refer the reader to the enclosed résumé for more detailed information.

Fourth Paragraph. In this paragraph say what you will do to follow up on your letter. For example, state that you will call by a certain date to set up an interview or to find out if the company will be recruiting in your area. Finish by indicating your willingness to answer any questions the recipient may have. Be sure you have provided your phone number.

Sincerely,

Type your name

Enclosure

Exhibit 2.4
SAMPLE COVER LETTER

15 Canal Street
Lockport, California 98772
(415) 555-6789
ldbernelli@xxx.com

September 7, 2006

Ms. Candace Kincaid
Director of Employment Services
California State Automobile Association
150 Vancouver Avenue
San Francisco, California 94102

Dear Ms. Kincaid:

Your opening for an actuarial analyst, advertised on The Monster Board–Job Details Web page, is of special interest to me. I graduated this past June with a bachelor's degree in mathematics from University College, and I would like to explore the possibility of employment with the California State Automobile Association.

The ad indicates that the job requires knowledge of statistical methods and insurance accounting concepts, along with strong PC skills. My résumé presents my work history in both statistical methods and accounting, and I have also added to my computer skills with an additional course in software for statisticians since graduation.

My work with the local chamber of commerce's Advertising Evaluation Project has given me an unusual degree of analytical practice for a variety of different businesses. In addition, I gained considerable experience in making regular presentations of my findings to the local merchants, which would allow me to represent the California State Automobile Association in an effective manner.

As you will see from my enclosed résumé, my college course work in my major was expanded and supported by advanced work in the areas of statistics and analysis. I am familiar with personal line insurance, and have studied all of the basic software programs and used state-of-the-art technology throughout college. I have completed one actuarial exam to date.

continued

I would like to meet with you to discuss how my background and abilities fit with your needs for this position. I will contact your office next week to discuss the possibility of an interview. In the meantime, if you have any questions or would like any additional information, please contact me at my home at (415) 555-6789. I look forward to talking with you.

Sincerely,

Lee D. Bernelli

Enclosure

3

Researching Careers
and Networking

One reason for confusion is perhaps a mistaken assumption that a college education provides job training. In most cases it does not. Of course, applied fields such as engineering, management, or education provide specific skills for the workplace as well as an education.

What Do They Call the Job You Want?

Your overall college education exposes you to numerous fields of study and teaches you quantitative reasoning, critical thinking, writing, and speaking, all of which can be successfully applied to a number of different job fields.

"What can I do with my math degree?" is a common question of many math majors who are approaching graduation and beginning to think about their postcollege jobs. It is likely that you have made some career explorations and you have a good idea of the general areas of applied mathematics that interest you most. This book will help you in researching careers further and focusing specifically on the careers that provide the best opportunity for you and your individual preferences and skills. It will help you answer questions such as "What career paths lead to management positions?" "What does a mathematician do in the insurance industry or in real estate?" "Where do mathematicians fit into the aeronautics industry, or in investment banking, or in a variety of government careers?"

But it still remains up to you to choose a job field and to learn how to articulate the benefits of your education in a way the employer will appreciate.

Collect Job Titles

The world of employment is a complex place, so you need to become a bit of an explorer and adventurer and be willing to try a variety of techniques to develop a list of possible occupations that might use your talents and education. You might find computerized interest inventories, reference books and other sources, and classified ads helpful in this respect. Once you have a list of possibilities that you are interested in and qualified for, you can move on to find out what kinds of organizations have these job titles.

Computerized Interest Inventories. One way to begin collecting job titles is to identify a number of jobs that call for your degree and the particular skills and interests you identified as part of the self-assessment process. There are excellent interactive career-guidance programs on the market to help you produce such selected lists of possible job titles. Most of these are available at colleges and at some larger town and city libraries. Two of the industry leaders are *CHOICES* and *DISCOVER*. Both allow you to enter interests, values, educational background, and other information to produce lists of possible occupations and industries. Each of the resources listed here will produce different job title lists. Some job titles will appear again and again, while others will be unique to a particular source. Investigate all of them!

Reference Sources. Books on the market that may be available through your local library or career counseling office also suggest various occupations related to specific majors. The following are only a few of the many good books on the market: *The College Board Guide to 150 Popular College Majors* and *College Majors and Careers: A Resource Guide for Effective Life Planning* both by Paul Phifer, and *Kaplan's What to Study: 101 Fields in a Flash.* All of these books list possible job titles within the academic major.

Finding an employer who needs someone with your skills is important, but equally important is how well the job and its environment fit your needs. Some employers and work environments will be much more attractive than others.

A math major considering a career as a statistician may work as a marketing/sales data analyst, a biostatistician in research, or a government statistician. Salaries could be exactly the same for each of these jobs, but each one probably presents a unique job culture. The expected norms of work pace, interactions, autonomy, and other factors can make a big difference in how well the job fits your needs.

If you have enjoyed analytical projects and developed strong skills, you may naturally be thinking of a job in statistics. However, math majors with these skills go on to work as retail buyers, actuaries, research analysts, administrators, and a variety of other jobs. Each of these jobs can be found in several different settings, and it is worthwhile to expand your awareness of the possibilities.

Each job title deserves your consideration. Like removing the layers of an onion, the search for job titles can go on and on! As you spend time doing this activity, you are actually learning more about the value of your degree. What's important in your search at this point is not to become critical or selective but rather to develop as long a list of possibilities as you can. Every source used will help you add new and potentially exciting jobs to your growing list.

Classified Ads. It has been well publicized that the classified ad section of the newspaper represents only a small fraction of the current job market. Nevertheless, the weekly classified ads can be a great help to you in your search. Although they may not be the best place to look for a job, they can teach you a lot about the job market. Classified ads provide a good education in job descriptions, duties, responsibilities, and qualifications. In addition, they provide insight into which industries are actively recruiting and some indication of the area's employment market. This is particularly helpful when seeking a position in a specific geographic area and/or a specific field. For your purposes, classified ads are a good source for job titles to add to your list.

Read the Sunday classified ads in a major market newspaper for several weeks in a row. Cut and paste all the ads that interest you and seem to call for something close to your education, skills, experience, and interests. Remember that classified ads are written for what an organization *hopes* to find; so you don't have to meet absolutely every criterion. However, if certain requirements are stated as absolute minimums and you cannot meet them, it's best not to waste your time and that of the employer.

The weekly classified want ads exercise is important because these jobs are out in the marketplace. They truly exist, and people with your qualifications are being sought to apply. What's more, many of these advertisements describe the duties and responsibilities of the job advertised and give you a beginning sense of the challenges and opportunities such a position presents. Some will indicate salary, and that will be helpful as well. This information will better define the jobs for you and provide some good material for possible interviews in that field.

Explore Job Descriptions

Once you've arrived at a solid list of possible job titles that interest you and for which you believe you are somewhat qualified, it's a good idea to do some research on each of these jobs. The preeminent source for such job information is the *Dictionary of Occupational Titles*, or *DOT* (wave.net/upg/immigration/dot_index.html). This directory lists every conceivable job and provides excellent up-to-date information on duties and responsibilities, interactions with associates, and day-to-day assignments and tasks. These descriptions provide a thorough job analysis, but they do not consider the possible employers or the environments in which a job may be performed. So, although a position as public relations officer may be well defined in terms of duties and responsibilities, it does not explain the differences in doing public relations work in a college or a hospital or a factory or a bank. You will need to look somewhere else for work settings.

Learn More About Possible Work Settings

After reading some job descriptions, you may choose to edit and revise your list of job titles once again, discarding those you feel are not suitable and keeping those that continue to hold your interest. Or you may wish to keep your list intact and see where these jobs may be located. For example, if you are interested in public relations and you appear to have those skills and the requisite education, you'll want to know which organizations do public relations. How can you find that out? How much income does someone in public relations make a year and what is the employment potential for the field of public relations?

To answer these and many other questions about your list of job titles, we recommend you try any of the following resources: *Careers Encyclopedia*, the professional societies and resources found throughout this book, *College to Career: The Guide to Job Opportunities*, and the *Occupational Outlook Handbook* (http://stats.bls.gov/ocohome.htm). Each of these resources, in a different way, will help to put the job titles you have selected into an employer context. Perhaps the most extensive discussion is found in the *Occupational Outlook Handbook*, which gives a thorough presentation of the nature of the work, the working conditions, employment statistics, training, other qualifications, and advancement possibilities as well as job outlook and earnings. Related occupations are also detailed, and a select bibliography is provided to help you find additional information.

Continuing with our public relations example, your search through these reference materials would teach you that the public relations jobs you find attractive are available in larger hospitals, financial institutions, most corporations (both consumer goods and industrial goods), media organizations, and colleges and universities.

Networking

Networking is the process of deliberately establishing relationships to get career-related information or to alert potential employers that you are available for work. Networking is critically important to today's job seeker for two reasons: it will help you get the information you need, and it can help you find out about *all* of the available jobs.

Get the Information You Need

Networkers will review your résumé and give you feedback on its effectiveness. They will talk about the job you are looking for and give you a candid appraisal of how they see your strengths and weaknesses. If they have a good sense of the industry or the employment sector for that job, you'll get their feelings on future trends in the industry as well. Some networkers will be very forthcoming about salaries, job-hunting techniques, and suggestions for your job search strategy. Many have been known to place calls right from the interview desk to friends and associates who might be interested in you. Each networker will make his or her own contribution, and each will be valuable.

Because organizations must evolve to adapt to current global market needs, the information provided by decision makers within various organizations will be critical to your success as a new job market entrant. For example, you might learn about the concept of virtual organizations from a networker. Virtual organizations coordinate economic activity to deliver value to customers by using resources outside the traditional boundaries of the organization. This concept is being discussed and implemented by chief executive officers of many organizations, including Ford Motor, Dell, and IBM. Networking can help you find out about this and other trends currently affecting the industries under your consideration.

Find Out About All of the Available Jobs

Not every job that is available at this very moment is advertised for potential applicants to see. This is called the *hidden job market*. Only 15 to 20 percent of all jobs are formally advertised, which means that 80 to 85 percent of available jobs do not appear in published channels. Networking will help you become more knowledgeable about all the employment opportunities available during your job search period.

Although someone you might talk to today doesn't know of any openings within his or her organization, tomorrow or next week or next month an opening may occur. If you've taken the time to show an interest in and knowledge of their organization, if you've shown the company representative how you can help achieve organizational goals and that you can fit into the organization, you'll be one of the first candidates considered for the position.

Networking: A Proactive Approach

Networking is a proactive rather than a reactive approach. You, as a job seeker, are expected to initiate a certain level of activity on your own behalf; you cannot afford to simply respond to jobs listed in the newspaper. Being proactive means building a network of contacts that includes informed and interested decision makers who will provide you with up-to-date knowledge of the current job market and increase your chances of finding out about employment opportunities appropriate for your interests, experience, and level of education. An old axiom of networking says, "You are only two phone calls away from the information you need." In other words, by talking to enough people, you will quickly come across someone who can offer you help.

Preparing to Network

In deliberately establishing relationships, maximize your efforts by organizing your approach. Five specific areas in which you can organize your efforts include reviewing your self-assessment, reviewing your research on job sites and organizations, deciding who you want to talk to, keeping track of all your efforts, and creating your self-promotion tools.

Review Your Self-Assessment

Your self-assessment is as important a tool in preparing to network as it has been in other aspects of your job search. You have carefully evaluated your personal traits, personal values, economic needs, longer-term goals, skill base, preferred skills, and underdeveloped skills. During the networking process you will be called upon to communicate what you know about yourself and relate it to the information or job you seek. Be sure to review the exercises that you completed in the self-assessment section of this book in preparation for networking. We've explained that you need to assess which skills you have acquired from your major that are of general value to an employer; be ready to express those in ways he or she can appreciate as useful in the organizations.

Review Research on Job Sites and Organizations

In addition, individuals assisting you will expect that you'll have at least some background information on the occupation or industry of interest to you. Refer to the appropriate sections of this book and other relevant publications to acquire the background information necessary for effective networking. They'll explain how to identify not only the job titles that might be of interest to you but also which kinds of organizations employ people to do that job. You will develop some sense of working conditions and expectations about duties and responsibilities—all of which will be of help in your networking interviews.

Decide Who You Want to Talk To

Networking cannot begin until you decide who you want to talk to and, in general, what type of information you hope to gain from your contacts. Once you know this, it's time to begin developing a list of contacts. Five useful sources for locating contacts are described here.

College Alumni Network. Most colleges and universities have created a formal network of alumni and friends of the institution who are particularly interested in helping currently enrolled students and graduates of their alma mater gain employment-related information.

It is usually a simple process to make use of an alumni network. Visit your college's website and locate the alumni office and/or your career center. Either or both sites will have information about your school's alumni network. You'll be provided with information on shadowing experiences, geographic information, or those alumni offering job referrals. If you don't find what you're looking for, don't hesitate to phone or e-mail your career center and ask what they can do to help you connect with an alum.

Alumni networkers may provide some combination of the following services: day-long shadowing experiences, telephone interviews, in-person interviews, information on relocating to given geographic areas, internship information, suggestions on graduate school study, and job vacancy notices.

Present and Former Supervisors. If you believe you are on good terms with present or former job supervisors, they may be an excellent resource for providing information or directing you to appropriate resources that would have information related to your current interests and needs. Additionally, these supervisors probably belong to professional organizations that they might be willing to utilize to get information for you.

Employers in Your Area. Although you may be interested in working in a geographic location different from the one where you currently reside, don't overlook the value of the knowledge and contacts those around you are able to provide. Use the local telephone directory and newspaper to identify the types of organizations you are thinking of working for or professionals who have the kinds of jobs you are interested in. Recently, a call made to a local hospital's financial administrator for information on working in health-care financial administration yielded more pertinent information on training seminars, regional professional organizations, and potential employment sites than a national organization was willing to provide.

Employers in Geographic Areas Where You Hope to Work. If you are thinking about relocating, identifying prospective employers or informational contacts in the new location will be critical to your success. Here are some tips for online searching. First, use a "metasearch" engine to get the most out of your search. Metasearch engines combine several engines into one powerful tool. We frequently use dogpile.com and metasearch.com for this purpose. Try using the city and state as your keywords in a search. *New Haven, Connecticut* will bring you to the city's website with links to the chamber of commerce, member businesses, and other valuable resources. By using looksmart.com you can locate newspapers in any area, and they, too, can provide valuable insight before you relocate. Of course, both dogpile and metasearch can lead you to yellow and white page directories in areas you are considering.

Professional Associations and Organizations. Professional associations and organizations can provide valuable information in several areas: career paths that you might not have considered, qualifications relating to those career choices, publications that list current job openings, and workshops or seminars that will enhance your professional knowledge and skills. They can also be excellent sources for background information on given industries: their health, current problems, and future challenges.

There are several excellent resources available to help you locate professional associations and organizations that would have information to meet your needs. Two especially useful publications are the *Encyclopedia of Associations* and *National Trade and Professional Associations of the United States*.

Keep Track of All Your Efforts

It can be difficult, almost impossible, to remember all the details related to each contact you make during the networking process, so you will want to develop a record-keeping system that works for you. Formalize this process by using your computer to keep a record of the people and organizations you want to contact. You can simply record the contact's name, address, and telephone number, and what information you hope to gain.

You could record this as a simple Word document and you could still use the "Find" function if you were trying to locate some data and could only recall the firm's name or the contact's name. If you're comfortable with database management and you have some database software on your computer, then you can put information at your fingertips even if you have only the zip code! The point here is not technological sophistication but good record keeping.

Once you have created this initial list, it will be helpful to keep more detailed information as you begin to actually make the contacts. Those details should include complete contact information, the date and content of each contact, names and information for additional networkers, and required follow-up. Don't forget to send a letter thanking your contact for his or her time! Your contact will appreciate your recall of details of your meetings and conversations, and the information will help you to focus your networking efforts.

Create Your Self-Promotion Tools

There are two types of promotional tools that are used in the networking process. The first is a résumé and cover letter, and the second is a one-minute "infomercial," which may be given over the telephone or in person.

Techniques for writing an effective résumé and cover letter are discussed in Chapter 2. Once you have reviewed that material and prepared these important documents, you will have created one of your self-promotion tools.

The one-minute infomercial will demand that you begin tying your interests, abilities, and skills to the people or organizations you want to network with. Think about your goal for making the contact to help you understand what you should say about yourself. You should be able to express yourself easily and convincingly. If, for example, you are contacting an alumnus of your institution to obtain the names of possible employment sites in a distant city, be prepared to discuss why you are interested in moving to that location, the types of jobs you are interested in, and the skills and abilities you possess that will make you a qualified candidate.

To create a meaningful one-minute infomercial, write it out, practice it as if it will be a spoken presentation, rewrite it, and practice it again if necessary until expressing yourself comes easily and is convincing.

Here's a simplified example of an infomercial for use over the telephone:

Hello, Mr. Hutter. My name is Monika Adair, and I was referred to you by Mr. Jerry Portillo of the chamber of commerce. I am a recent graduate of South-East State University, and I want to enter the investment field. I have a bachelor of science degree in mathematics, with a strong quantitative background and good research and computer skills.

I· am calling to request an information interview. I need more information about the investment field and where I might fit in. I am hoping that you would be willing to meet with me, for about a half hour, to discuss your perspectives on careers in investment counseling. There are so many possible employers to approach, and I am seeking advice on which ones might be the best possibilities for my particular skills and experience.

Would you be willing to discuss this with me? I would greatly appreciate it. I am available most mornings if that would be convenient for you.

It very well may happen that your employer contact wishes you to communicate by e-mail. The infomercial quoted above could easily be rewritten for an e-mail message. You should "cut and paste" your résumé right into the e-mail text itself.

Other effective self-promotion tools include portfolios for those in the arts, writing professions, or teaching. Portfolios show examples of work, photographs of projects or classroom activities, or certificates and credentials that are job related. There may not be an opportunity to use the portfolio during an interview, and it is not something that should be left with the organization. It is designed to be explained and displayed by the creator. However, during some networking meetings, there may be an opportunity to illustrate a point or strengthen a qualification by exhibiting the portfolio.

Beginning the Networking Process

Set the Tone for Your Communications

It can be useful to establish "tone words" for any communications you embark upon. Before making your first telephone call or writing your first letter, decide what you want the person to think of you. If you are networking to try to obtain a job, your tone words might include descriptors such as *genuine*, *informed*, and *self-knowledgeable*. When you're trying to acquire information, your tone words may have a slightly different focus, such as *courteous*, *organized*, *focused*, and *well-spoken*. Use the tone words you establish for your contacts to guide you through the networking process.

Honestly Express Your Intentions

When contacting individuals, it is important to be honest about your reasons for making the contact. Establish your purpose in your own mind and be able and ready to articulate it concisely. Determine an initial agenda, whether it be informational questioning or self-promotion, present it to your contact, and be ready to respond immediately. If you don't adequately prepare before initiating your overture, you may find yourself at a disadvantage if you're asked to immediately begin your informational interview or self-promotion during the first phone conversation or visit.

Start Networking Within Your Circle of Confidence

Once you have organized your approach—by utilizing specific researching methods, creating a system for keeping track of the people you will contact, and developing effective self-promotion tools—you are ready to begin networking. The best way to begin networking is by talking with a group of people you trust and feel comfortable with. This group is usually made up of your family, friends, and career counselors. No matter who is in this inner circle, they will have a special interest in seeing you succeed in your job search. In addition, because they will be easy to talk to, you should try taking some risks in terms of practicing your information-seeking approach. Gain confidence in talking about the strengths you bring to an organization and the underdeveloped skills you feel hinder your candidacy. Be sure to review the section on self-assessment for tips on approaching each of these areas. Ask for critical but constructive feedback from the people in your circle of confidence on the letters you write and the one-minute infomercial you have developed. Evaluate whether you want to make the changes they suggest, then practice the changes on others within this circle.

Stretch the Boundaries of Your Networking Circle of Confidence

Once you have refined the promotional tools you will use to accomplish your networking goals, you will want to make additional contacts. Because you will not know most of these people, it will be a less comfortable activity to undertake. The practice that you gained with your inner circle of trusted friends should have prepared you to now move outside of that comfort zone.

It is said that any information a person needs is only two phone calls away, but the information cannot be gained until you (1) make a reasonable guess about who might have the information you need and (2) pick up the telephone to make the call. Using your network list that includes alumni, instructors, supervisors, employers, and associations, you can begin preparing your list of questions that will allow you to get the information you need.

Prepare the Questions You Want to Ask

Networkers can provide you with the insider's perspective on any given field and you can ask them questions that you might not want to ask in an interview. For example, you can ask them to describe the more repetitious or mundane parts of the job or ask them for a realistic idea of salary expectations. Be sure to prepare your questions ahead of time so that you are organized and efficient.

Be Prepared to Answer Some Questions

To communicate effectively, you must anticipate questions that will be asked of you by the networkers you contact. Revisit the self-assessment process you undertook and the research you've done so that you can effortlessly respond to questions about your short- and long-term goals and the kinds of jobs you are most interested in pursuing.

General Networking Tips

Make Every Contact Count. Setting the tone for each interaction is critical. Approaches that will help you communicate in an effective way include politeness, being appreciative of time provided to you, and being prepared and thorough. Remember, *everyone* within an organization has a circle of influence, so be prepared to interact effectively with each person you encounter in the networking process, including secretarial and support staff. Many information or job seekers have thwarted their own efforts by being rude to some individuals they encountered as they networked because they made the incorrect assumption that certain persons were unimportant.

Sometimes your contacts may be surprised at their ability to help you. After meeting and talking with you, they might think they have not offered much in the way of help. A day or two later, however, they may make a contact that would be useful to you and refer you to that person.

With Each Contact, Widen Your Circle of Networkers. Always leave an informational interview with the names of at least two more people who can help you get the information or job that you are seeking. Don't be shy about asking for additional contacts; networking is all about increasing the number of people you can interact with to achieve your goals.

Make Your Own Decisions. As you talk with different people and get answers to the questions you pose, you may hear conflicting information or get conflicting suggestions. Your job is to listen to these "experts" and decide what information and which suggestions will help you achieve *your* goals. Only implement those suggestions that you believe will work for you.

Shutting Down Your Network

As you achieve the goals that motivated your networking activity—getting the information you need or the job you want—the time will come to inactivate all or parts of your network. As you do, be sure to tell your primary supporters about your change in status. Call or write to each one of them and give them as many details about your new status as you feel is necessary to maintain a positive relationship.

Because a network takes on a life of its own, activity undertaken on your behalf will continue even after you cease your efforts. As you get calls or are contacted in some fashion, be sure to inform these networkers about your change in status, and thank them for assistance they have provided.

Information on the latest employment trends indicates that workers will change jobs or careers several times in their lifetime. Networking, then, will be a critical aspect in the span of your professional life. If you carefully and thoughtfully conduct your networking activities during your job search, you will have a solid foundation of experience when you need to network the next time around.

Where Are These Jobs, Anyway?

Having a list of job titles that you've designed around your own career interests and skills is an excellent beginning. It means you've really thought about who you are and what you are presenting to the employment market. It has caused you to think seriously about the most appealing environments to work in, and you have identified some employer types that represent these environments.

The research and the thinking that you've done thus far will be used again and again. They will be helpful in writing your résumé and cover letters, in talking about yourself on the telephone to prospective employers, and in answering interview questions.

Now is a good time to begin to narrow the field of job titles and employment sites down to some specific employers to initiate the employment contact.

Find Out Which Employers Hire People Like You

This section will provide tips, techniques, and specific resources for developing an actual list of specific employers that can be used to make contacts. It is only an outline that you must be prepared to tailor to your own particular needs and according to what you bring to the job search. Once again, it is important to communicate with others along the way exactly what you're looking for and what your goals are for the research you're doing. Librarians, employers, career counselors, friends, friends of friends, business contacts, and bookstore staff will all have helpful information on geographically specific and new resources to aid you in locating employers who'll hire you.

Identify Information Resources

Your interview wardrobe and your new résumé might have put a dent in your wallet, but the resources you'll need to pursue your job search are available for free. The categories of information detailed here are not hard to find and are yours for the browsing.

Numerous resources described in this section will help you identify actual employers. Use all of them or any others that you identify as available in your geographic area. As you become experienced in this process, you'll quickly figure out which information sources are helpful and which are not. If you live in a rural area, a well-planned day trip to a major city that includes a college career office, a large college or city library, state and federal employment centers, a chamber of commerce office, and a well-stocked bookstore can produce valuable results.

There are many excellent resources available to help you identify actual job sites. They are categorized into employer directories (usually indexed by product lines and geographic location), geographically based directories (designed to highlight particular cities, regions, or states), career-specific directories (e.g., *Sports MarketPlace*, which lists tens of thousands of firms involved with sports), periodicals and newspapers, targeted job posting publications, and videos. This is by no means meant to be a complete treatment of resources but rather a starting point for identifying useful resources.

Working from the more general references to highly specific resources, we provide a basic list to help you begin your search. Many of these you'll find easily available. In some cases reference librarians and others will suggest even better materials for your particular situation. Start to create your own customized bibliography of job search references.

Geographically Based Directories. The Job Bank series published by Bob Adams, Inc. (aip.com) contains detailed entries on each area's major employers, including business activity, address, phone number, and hiring contact name. Many listings specify educational backgrounds being sought in potential employees. Each volume contains a solid discussion of each city's or state's major employment sectors. Organizations are also indexed by industry. Job Bank volumes are available for the following places: Atlanta, Boston, Chicago, Dallas–Ft. Worth, Denver, Detroit, Florida, Houston, Los Angeles, Minneapolis, New York, Ohio, Philadelphia, San Francisco, Seattle, St. Louis, Washington, D.C., and other cities throughout the Northwest.

National Job Bank (careercity.com) lists employers in every state, along with contact names and commonly hired job categories. Included are many small companies often overlooked by other directories. Companies are also indexed by industry. This publication provides information on educational backgrounds sought and lists company benefits.

Periodicals and Newspapers. Several sources are available to help you locate which journals or magazines carry job advertisements in your field. Other resources help you identify opportunities in other parts of the country.

- *Where the Jobs Are: A Comprehensive Directory of 1200 Journals Listing Career Opportunities*
- *Corptech Fast 5000 Company Locator*
- *National Ad Search* (nationaladsearch.com)
- *The Federal Jobs Digest* (jobsfed.com) and *Federal Career Opportunities*
- *World Chamber of Commerce Directory* (chamberofcommerce.org)

This list is certainly not exhaustive; use it to begin your job search work.

Targeted Job Posting Publications. Although the resources that follow are national in scope, they are either targeted to one medium of contact (telephone), focused on specific types of jobs, or less comprehensive than the sources previously listed.

- Careers.org (careers.org/index.html)
- *The Job Hunter* (jobhunter.com)
- *Current Jobs for Graduates* (graduatejobs.com)
- *Environmental Opportunities* (ecojobs.com)
- *Y National Vacancy List* (ymca.net/employment/ymca_recruiting/jobright.htm)
- *ArtSEARCH*
- *Community Jobs*
- *National Association of Colleges and Employers: Job Choices series*
- *National Association of Colleges and Employers* (jobweb.com)

Videos. You may be one of the many job seekers who likes to get information via a medium other than paper. Many career libraries, public libraries, and career centers in libraries carry an assortment of videos that will help you learn new techniques and get information helpful in the job search.

Locate Information Resources

Throughout these introductory chapters, we have continually referred you to various websites for information on everything from job listings to career information. Using the Web gives you a mobility at your computer that you don't enjoy if you rely solely on books or newspapers or printed journals. Moreover, material on the Web, if the site is maintained, can be the most up-to-date information available.

You'll eventually identify the information resources that work best for you, but make certain you've covered the full range of resources before you begin to rely on a smaller list. Here's a short list of informational sites that many job seekers find helpful:

- Public and college libraries
- College career centers
- Bookstores
- The Internet
- Local and state government personnel offices
- Career/job fairs

Each one of these sites offers a collection of resources that will help you get the information you need.

As you meet and talk with service professionals at all these sites, be sure to let them know what you're doing. Inform them of your job search, what you've already accomplished, and what you're looking for. The more people who know you're job seeking, the greater the possibility that someone will have information or know someone who can help you along your way.

4

Interviewing and Job Offer Considerations

Certainly, there can be no one part of the job search process more fraught with anxiety and worry than the interview. Yet seasoned job seekers welcome the interview and will often say, "Just get me an interview and I'm on my way!" They understand that the interview is crucial to the hiring process and equally crucial for them, as job candidates, to have the opportunity of a personal dialogue to add to what the employer may already have learned from the résumé, cover letter, and telephone conversations.

Believe it or not, the interview is to be welcomed, and even enjoyed! It is a perfect opportunity for you, the candidate, to sit down with an employer and express yourself and display who you are and what you want. Of course, it takes thought and planning and a little strategy; after all, it *is* a job interview! But it can be a positive, if not pleasant, experience and one you can look back on and feel confident about your performance and effort.

For many new job seekers, a job, any job, seems a wonderful thing. But seasoned interview veterans know that the job interview is an important step for both sides—the employer and the candidate—to see what each has to offer and whether there is going to be a "fit" of personalities, work styles, and attitudes. And it is this concept of balance in the interview, that both sides have important parts to play, that holds the key to success in mastering this aspect of the job search strategy.

Try to think of the interview as a conversation between two interested and equal partners. You both have important, even vital, information to deliver and to learn. Of course, there's no denying the employer has some leverage, especially in the initial interview for recruitment or any interview scheduled by the candidate and not the recruiter. That should not prevent the interviewee from seeking to play an equal part in what should be a fair exchange of information. Too often the untutored candidate allows the interview to become one-sided. The employer asks all the questions and the candidate simply responds. The ideal would be for two mutually interested parties to sit down and discuss possibilities for each. This is a conversation of significance, and it requires preparation, thought about the tone of the interview, and planning of the nature and details of the information to be exchanged.

Preparing for the Interview

The length of most initial interviews is about thirty minutes. Given the brevity, the information that is exchanged ought to be important. The candidate should be delivering material that the employer cannot discover on the résumé, and in turn, the candidate should be learning things about the employer that he or she could not otherwise find out. After all, if you have only thirty minutes, why waste time on information that is already published? The information exchanged is more than just factual, and both sides will learn much from what they see of each other, as well. How the candidate looks, speaks, and acts are important to the employer. The employer's attention to the interview and awareness of the candidate's résumé, the setting, and the quality of information presented are important to the candidate.

Just as the employer has every right to be disappointed when a prospect is late for the interview, looks unkempt, and seems ill-prepared to answer fairly standard questions, the candidate may be disappointed with an interviewer who isn't ready for the meeting, hasn't learned the basic résumé facts, and is constantly interrupted by telephone calls. In either situation there's good reason to feel let down.

There are many elements to a successful interview, and some of them are not easy to describe or prepare for. Sometimes there is just a chemistry between interviewer and interviewee that brings out the best in both, and a good exchange takes place. But there is much the candidate can do to pave the way for success in terms of his or her résumé, personal appearance, goals, and interview strategy—each of which we will discuss. However, none of this preparation is as important as the time and thought the candidate gives to personal self-assessment.

Self-Assessment

Neither a stunning résumé nor an expensive, well-tailored suit can compensate for candidates who do not know what they want, where they are going, or why they are interviewing with a particular employer. Self-assessment, the process by which we begin to know and acknowledge our own particular blend of education, experiences, needs, and goals, is not something that can be sorted out the weekend before a major interview. Of all the elements of interview preparation, this one requires the longest lead time and cannot be faked.

Because the time allotted for most interviews is brief, it is all the more important for job candidates to understand and express succinctly why they are there and what they have to offer. This is not a time for undue modesty (or for braggadocio either); it is a time for a compelling, reasoned statement of why you feel that you and this employer might make a good match. It means you have to have thought about your skills, interests, and attributes; related those to your life experiences and your own history of challenges and opportunities; and determined what that indicates about your strengths, preferences, values, and areas needing further development.

If you need some assistance with self-assessment issues, refer to Chapter 1. Included are suggested exercises that can be done as needed, such as making up an experiential diary and extracting obvious strengths and weaknesses from past experiences. These simple assignments will help you look at past activities as collections of tasks with accompanying skills and responsibilities. Don't overlook your high school or college career office. Many offer personal counseling on self-assessment issues and may provide testing instruments such as the *Myers-Briggs Type Indicator (MBTI)*, the *Harrington-O'Shea Career Decision-Making System (CDM)*, the *Strong Interest Inventory (SII)*, or any other of a wide selection of assessment tools that can help you clarify some of these issues prior to the interview stage of your job search.

The Résumé

Résumé preparation has been discussed in detail, and some basic examples were provided. In this section we want to concentrate on how best to use your résumé in the interview. In most cases the employer will have seen the résumé prior to the interview, and, in fact, it may well have been the quality of that résumé that secured the interview opportunity.

An interview is a conversation, however, and not an exercise in reading. So, if the employer hasn't seen your résumé and you have brought it along to the interview, wait until asked or until the end of the interview to offer it. Otherwise, you may find yourself staring at the back of your résumé and simply answering "yes" and "no" to a series of questions drawn from that document.

Sometimes an interviewer is not prepared and does not know or recall the contents of the résumé and may use the résumé to a greater or lesser degree as a "prompt" during the interview. It is for you to judge what that may indicate about the individual performing the interview or the employer. If your interviewer seems surprised by the scheduled meeting, relies on the résumé to an inordinate degree, and seems otherwise unfamiliar with your background, this lack of preparation for the hiring process could well be a symptom of general management disorganization or may simply be the result of poor planning on the part of one individual. It is your responsibility as a potential employee to be aware of these signals and make your decisions accordingly.

If the interviewer is preoccupied with your résumé, you may be losing an important opportunity to discuss your potential. You can politely bring the interview back to an interactive mode by saying something like, "Mr. Conley, you may be interested in some very recent experience I have gained as a volunteer in working with a nonprofit organization's budget analysis and forecast. It is not detailed on my résumé. May I tell you about it?"

By all means, bring at least one copy of your résumé to the interview. Occasionally, at the close of an interview, an interviewer will express an interest in circulating a résumé to several departments, and you could then offer the copy you brought. Sometimes, an interview appointment provides an opportunity to meet others in the organization who may express an interest in you and your background, and it may be helpful to follow up with a copy of your résumé. Our best advice, however, is to keep it out of sight until needed or requested.

Employer Information

Whether your interview is for graduate school admission, an overseas corporate position, or a position with a local company, it is important to know something about the employer or the organization. Keeping in mind that the interview is relatively brief and that you will hopefully have other interviews with other organizations, it is important to keep your research in proportion. If secondary interviews are called for, you will have additional time to do further research. For the first interview, it is helpful to know the organization's mission, goals, size, scope of operations, and so forth. Your research may uncover recent areas of challenge or particular successes that may help to fuel the interview. Use the "What Do They Call the Job You Want?" section of Chapter 3, your library, and your career or guidance office to help you locate this information in the most efficient way possible. Don't be shy in asking advice of these counseling and guidance professionals on how best to spend your preparation time. With some practice, you'll soon learn how much information is enough and which kinds of information are most useful to you.

Interview Content

We've already discussed how it can help to think of the interview as an important conversation—one that, as with any conversation, you want to find pleasant and interesting and to leave you with a good feeling. But because this conversation is especially important, the information that's exchanged is critical to its success. What do you want them to know about you? What do you need to know about them? What interview technique do you need to particularly pay attention to? How do you want to manage the close of the interview? What steps will follow in the hiring process?

Except for the professional interviewer, most of us find interviewing stressful and anxiety-provoking. Developing a strategy before you begin interviewing will help you relieve some stress and anxiety. One particular strategy that has worked for many and may work for you is interviewing by objective. Before you interview, write down three to five goals you would like to achieve for that interview. They may be technique goals: smile a little more, have a firmer handshake, be sure to ask about the next stage in the interview process before leaving. They may be content-oriented goals: find out about the company's current challenges and opportunities; be sure to speak of your recent research, writing experiences, or foreign travel. Whatever your goals, jot down a few of them as goals for each interview.

Most people find that in trying to achieve these few goals, their interviewing technique becomes more organized and focused. After the interview, the most common question friends and family ask is "How did it go?" With this technique, you have an indication of whether you met *your* goals for the meeting, not just some vague idea of how it went. Chances are, if you accomplished what you wanted to, it improved the quality of the entire interview. As you continue to interview, you will want to revise your goals to continue improving your interview skills.

Now, add to the concept of the significant conversation the idea of a beginning, a middle, and a closing and you will have two thoughts that will give your interview a distinctive character. Be sure to make your introduction warm and cordial. Say your full name (and if it's a difficult-to-pronounce name, help the interviewer to pronounce it) and make certain you know your interviewer's name and how to pronounce it. Most interviews begin with some "soft talk" about the weather, chat about the candidate's trip to the interview site, or national events. This is done as a courtesy to relax both you and the interviewer, to get you talking, and to generally try to defuse the atmosphere of excessive tension. Try to be yourself, engage in the conversation, and don't try to second-guess the interviewer. This is simply what it appears to be— casual conversation.

Once you and the interviewer move on to exchange more serious information in the middle part of the interview, the two most important concerns become your ability to handle challenging questions and your success at asking meaningful ones. Interviewer questions will probably fall into one of three categories: personal assessment and career direction, academic assessment, and knowledge of the employer. Here are a few examples of questions in each category:

Personal Assessment and Career Direction
1. What motivates you to put forth your best effort?
2. What do you consider to be your greatest strengths and weaknesses?
3. What qualifications do you have that make you think you will be successful in this career?

Academic Assessment
1. What led you to choose your major?
2. What subjects did you like best and least? Why?
3. How has your college experience prepared you for this career?

Knowledge of the Employer
1. What do you think it takes to be successful in an organization like ours?
2. In what ways do you think you can make a contribution to our organization?
3. Why did you choose to seek a position with this organization?

The interviewer wants a response to each question but is also gauging your enthusiasm, preparedness, and willingness to communicate. In each response you should provide some information about yourself that can be related to the employer's needs. A common mistake is to give too much information. Answer each question completely, but be careful not to run on too long with extensive details or examples.

Questions About Underdeveloped Skills

Most employers interview people who have met some minimum criteria of education and experience. They interview candidates to see who they are, to learn what kind of personality they exhibit, and to get some sense of how they might fit into the existing organization. It may be that you are asked about skills the employer hopes to find and that you have not documented. Maybe it's grant-writing experience, knowledge of the European political system, or a knowledge of the film world.

To questions about skills and experiences you don't have, answer honestly and forthrightly and try to offer some additional information about skills you do have. For example, perhaps the employer is disappointed you have no grant-writing experience. An honest answer may be as follows:

> *No, unfortunately, I was never in a position to acquire those skills. I do understand something of the complexities of the grant-writing process and feel confident that my attention to detail, careful reading skills, and strong writing would make grants a wonderful challenge in a new job. I think I could get up on the learning curve quickly.*

The employer hears an honest admission of lack of experience but is reassured by some specific skill details that do relate to grant writing and a confident manner that suggests enthusiasm and interest in a challenge.

For many students, questions about their possible contribution to an employer's organization can prove challenging. Because your education has probably not included specific training for a job, you need to review your academic record and select capabilities you have developed in your major that an employer can appreciate. For example, perhaps you read well and can analyze and condense what you've read into smaller, more focused pieces. That could be valuable. Or maybe you did some serious research and you know you have valuable investigative skills. Your public speaking might be highly developed and you might use visual aids appropriately and effectively. Or maybe your skill at correspondence, memos, and messages is effective. Whatever it is, you must take it out of the academic context and put it into a new, employer-friendly context so your interviewer can best judge how you could help the organization.

Exhibiting knowledge of the organization will, without a doubt, show the interviewer that you are interested enough in the available position to have done some legwork in preparation for the interview. Remember, it is not necessary to know every detail of the organization's history but rather to have a general knowledge about why it is in business and how the industry is faring.

Sometime during the interview, generally after the midway point, you'll be asked if you have any questions for the interviewer. Your questions will tell the employer much about your attitude and your desire to understand the organization's expectations so you can compare them to your own strengths. The following are just a few questions you might want to ask:

1. What is the communication style of the organization? (meetings, memos, and so forth)
2. What would a typical day in this position be like for me?
3. What have been some of the interesting challenges and opportunities your organization has recently faced?

Most interviews draw to a natural closing point, so be careful not to prolong the discussion. At a signal from the interviewer, wind up your presentation, express your appreciation for the opportunity, and be sure to ask what the next stage in the process will be. When can you expect to hear from them? Will they be conducting second-tier interviews? If you are interested and haven't heard, would they mind a phone call? Be sure to collect a business card with the name and phone number of your interviewer. On your way out, you might have an opportunity to pick up organizational literature you haven't seen before.

With the right preparation—a thorough self-assessment, professional clothing, and employer information—you'll be able to set and achieve the goals you have established for the interview process.

Interview Follow-Up

Quite often there is a considerable time lag between interviewing for a position and being hired or, in the case of the networker, between your phone call or letter to a possible contact and the opportunity of a meeting. This can be frustrating. "Why aren't they contacting me?" "I thought I'd get another interview, but no one has telephoned." "Am I out of the running?" You don't know what is happening.

Consider the Differing Perspectives

Of course, there is another perspective—that of the networker or hiring organization. Organizations are complex, with multiple tasks that need to be accomplished each day. Hiring is a discrete activity that does not occur as frequently as other job assignments. The hiring process might have to take second place to other, more immediate organizational needs. Although it may be very important to you, and it is certainly ultimately significant to the employer, other issues such as fiscal management, planning and product development, employer vacation periods, or financial constraints may prevent an organization or individual within that organization from acting on your employment or your request for information as quickly as you or they would prefer.

Use Your Communication Skills

Good communication is essential here to resolve any anxieties, and the responsibility is on you, the job or information seeker. Too many job seekers and networkers offer as an excuse that they don't want to "bother" the organization by writing letters or calling. Let us assure you here and now, once and for all, that if you are troubling an organization by over-communicating, someone will indicate that situation to you quite clearly. If not, you can only assume you are a worthwhile prospect and the employer appreciates being reminded of your availability and interest. Let's look at follow-up practices in the job interview process and the networking situation separately.

Following Up on the Employment Interview

A brief thank-you note following an interview is an excellent and polite way to begin a series of follow-up communications with a potential employer with whom you have interviewed and want to remain in touch. It should be just that—a thank-you for a good meeting. If you failed to mention some fact or experience during your interview that you think might add to your candidacy, you may use this note to do that. However, this should be essentially a note whose overall tone is appreciative and, if appropriate, indicative of a continuing interest in pursuing any opportunity that may exist with that organization. It is one of the few pieces of business correspondence that may be handwritten, but always use plain, good-quality, standard-size paper.

If, however, at this point you are no longer interested in the employer, the thank-you note is an appropriate time to indicate that. You are under no obligation to identify any reason for not continuing to pursue employment with that organization, but if you are so inclined to indicate your professional reasons (pursuing other employers more akin to your interests, looking for greater income production than this employer can provide, a different geographic location), you certainly may. It should not be written with an eye to negotiation, for it will not be interpreted as such.

As part of your interview closing, you should have taken the initiative to establish lines of communication for continuing information about your candidacy. If you asked permission to telephone, wait a week following your thank-you note, then telephone your contact simply to inquire how things are progressing on your employment status. The feedback you receive here should be taken at face value. If your interviewer simply has no information, he or she will tell you so and indicate whether you should call again and when. Don't be discouraged if this should continue over some period of time.

If during this time something occurs that you think improves or changes your candidacy (some new qualification or experience you may have had), including any offers from other organizations, by all means telephone or write to inform the employer about this. In the case of an offer from a competing but less desirable or equally desirable organization, telephone your contact, explain what has happened, express your real interest in the organization, and inquire whether some determination on your employment might be made before you must respond to this other offer. An organization that is truly interested in you may be moved to make a decision about your candidacy. Equally possible is the scenario in which they are not yet ready to make a decision and so advise you to take the offer that has been presented. Again, you have no ethical alternative but to deal with the information presented in a straightforward manner.

When accepting other employment, be sure to contact any employers still actively considering you and inform them of your new job. Thank them graciously for their consideration. There are many other job seekers out there just like you who will benefit from having their candidacy improved when others bow out of the race. Who knows, you might at some future time have occasion to interact professionally with one of the organizations with which you sought employment. How embarrassing it would be to have someone remember you as the candidate who failed to notify them that you were taking a job elsewhere!

In all of your follow-up communications, keep good notes of whom you spoke with, when you called, and any instructions that were given about return communications. This will prevent any misunderstandings and provide you with good records of what has transpired.

Job Offer Considerations

For many recent college graduates, the thrill of their first job and, for some, the most substantial regular income they have ever earned seems an excess of good fortune coming at once. To question that first income or to be critical in any way of the conditions of employment at the time of the initial offer seems like looking a gift horse in the mouth. It doesn't seem to occur to many new hires even to attempt to negotiate any aspect of their first job. And, as many employers who deal with entry-level jobs for recent college graduates will readily confirm, the reality is that there simply isn't much movement in salary available to these new college recruits. The entry-level hire generally does not have an employment track record on a professional level to provide any leverage for negotiation. Real negotiations on salary, benefits, retirement provisions, and so forth come to those with significant employment records at higher income levels.

Of course, the job offer is more than just money. It can be composed of geographic assignment, duties and responsibilities, training, benefits, health and medical insurance, educational assistance, car allowance or company vehicle, and a host of other items. All of this is generally detailed in the formal letter that presents the final job offer. In most cases this is a follow-up to a personal phone call from the employer representative who has been principally responsible for your hiring process.

That initial telephone offer is certainly binding as a verbal agreement, but most firms follow up with a detailed letter outlining the most significant parts of your employment contract. You may, of course, choose to respond immediately at the time of the telephone offer (which would be considered a binding oral contract), but you will also be required to formally answer the letter of offer with a letter of acceptance, restating the salient elements of the employer's description of your position, salary, and benefits. This ensures that both parties are clear on the terms and conditions of employment and remuneration and any other outstanding aspects of the job offer.

Is This the Job You Want?

Most new employees will respond affirmatively in writing, glad to be in the position to accept employment. If you've worked hard to get the offer and the job market is tight, other offers may not be in sight, so you will say, "Yes, I accept!" What is important here is that the job offer you accept be one that does fit your particular needs, values, and interests as you've outlined them in your self-assessment process. Moreover, it should be a job that will not only use your skills and education but also challenge you to develop new skills and talents.

Jobs are sometimes accepted too hastily, for the wrong reasons, and without proper scrutiny by the applicant. For example, an individual might readily accept a sales job only to find the continual rejection by potential clients unendurable. An office worker might realize within weeks the constraints of a desk job and yearn for more activity. Employment is an important part of our lives. It is, for most of our adult lives, our most continuous productive activity. We want to make good choices based on the right criteria.

If you have a low tolerance for risk, a job based on commission will certainly be very anxiety-provoking. If being near your family is important, issues of relocation could present a decision crisis for you. If you're an adventurous person, a job with frequent travel would provide needed excitement and be very desirable. The importance of income, the need to continue your education, your personal health situation—all of these have an impact on whether the job you are considering will ultimately meet your needs. Unless you've spent some time understanding and thinking about these issues, it will be difficult to evaluate offers you do receive.

More important, if you make a decision that you cannot tolerate and feel you must leave that job, you will then have both unemployment and self-esteem issues to contend with. These will combine to make the next job search tough going, indeed. So make your acceptance a carefully considered decision.

Negotiate Your Offer

It may be that there is some aspect of your job offer that is not particularly attractive to you. Perhaps there is no relocation allotment to help you move your possessions, and this presents some financial hardship for you. It may be that the health insurance is less than you had hoped. Your initial assignment may be different from what you expected, either in its location or in the duties and responsibilities that comprise it. Or it may simply be that the salary is less than you anticipated. Other considerations may be your official starting date of employment, vacation time, evening hours, dates of training programs or schools, and other concerns.

If you are considering not accepting the job because of some item or items in the job offer "package" that do not meet your needs, you should know that most employers emphatically wish that you would bring that issue to their attention. It may be that the employer can alter it to make the offer more agreeable for you. In some cases it cannot be changed. In any event the employer would generally like to have the opportunity to try to remedy a difficulty rather than risk losing a good potential employee over an issue that might have been resolved. After all, they have spent time and funds in securing your services, and they certainly deserve an opportunity to resolve any possible differences.

Honesty is the best approach in discussing any objections or uneasiness you might have over the employer's offer. Having received your formal offer in writing, contact your employer representative and indicate your particular dissatisfaction in a straightforward manner. For example, you might explain that while you are very interested in being employed by this organization, the salary (or any other benefit) is less than you have determined you require. State the terms you need, and listen to the response. You may be asked to put this in writing, or you may be asked to hold off until the firm can decide on a response. If you are dealing with a senior representative of the organization, one who has been involved in hiring for some time, you may get an immediate response or a solid indication of possible outcomes.

Perhaps the issue is one of relocation. Your initial assignment is in the Midwest, and because you had indicated a strong West Coast preference, you are surprised at the actual assignment. You might simply indicate that while you understand the need for the company to assign you based on its needs, you are disappointed and had hoped to be placed on the West Coast. You could inquire if that were still possible and, if not, would it be reasonable to expect a West Coast relocation in the future.

If your request is presented in a reasonable way, most employers will not see this as jeopardizing your offer. If they can agree to your proposal, they will. If not, they will simply tell you so, and you may choose to continue your candidacy with them or remove yourself from consideration. The choice will be up to you.

Some firms will adjust benefits within their parameters to meet the candidate's need if at all possible. If a candidate requires a relocation cost allowance, he or she may be asked to forgo tuition benefits for the first year to accomplish this adjustment. An increase in life insurance may be adjusted by some other benefit trade-off; perhaps a family dental plan is not needed. In these decisions you are called upon, sometimes under time pressure, to know how you value these issues and how important each is to you.

Many employers find they are more comfortable negotiating for candidates who have unique qualifications or who bring especially needed expertise to the organization. Employers hiring large numbers of entry-level college graduates may be far more reluctant to accommodate any changes in offer conditions. They are well supplied with candidates with similar education and experience so that if rejected by one candidate, they can draw new candidates from an ample labor pool.

Compare Offers

The condition of the economy, the job seeker's academic major and particular geographic job market, and individual needs and demands for certain employment conditions may not provide more than one job offer at a time. Some job seekers may feel that no reasonable offer should go unaccepted for the simple fear there won't be another.

In a tough job market, or if the job you seek is not widely available, or when your job search goes on too long and becomes difficult to sustain financially and emotionally, it may be necessary to accept an inferior offer. The alternative is continued unemployment. Even here, when you feel you don't have a choice, you can at least understand that in accepting this particular offer, there may be limitations and conditions you don't appreciate. At the time of acceptance, there were no other alternatives, but you can begin to use that position to gain the experience and talent to move toward a more attractive position.

Sometimes, however, more than one offer is received, and the candidate has the luxury of choice. If the job seeker knows what he or she wants and has done the necessary self-assessment honestly and thoroughly, it may be clear that one of the offers conforms more closely to those expressed wants and needs.

However, if, as so often happens, the offers are similar in terms of conditions and salary, the question then becomes which organization might provide the necessary climate, opportunities, and advantages for your professional development and growth. This is the time when solid employer research and astute questioning during the interviews really pay off. How much did you learn about the employer through your own research and skillful questioning? When the interviewer asked during the interview "Do you have any questions?" did you ask the kinds of questions that would help resolve a choice between one organization and another? Just as an employer must decide among numerous applicants, so must the applicant learn to assess the potential employer. Both are partners in the job search.

Reneging on an Offer
An especially disturbing occurrence for employers and career counseling professionals is when a job seeker formally (either orally or by written contract) accepts employment with one organization and later reneges on the agreement and goes with another employer.

There are all kinds of rationalizations offered for this unethical behavior. None of them satisfies. The sad irony is that what the job seeker is willing to do to the employer—make a promise and then break it—he or she would be outraged to have done to him- or herself: have the job offer pulled. It is a very bad way to begin a career. It suggests the individual has not taken the time to do the necessary self-assessment and self-awareness exercises to think and judge critically. The new offer taken may, in fact, be no better or worse than the one refused. You should be aware that there have been incidents of legal action following job candidates' reneging on an offer. This adds a very sour note to what should be a harmonious beginning of a lifelong adventure.

PART TWO

THE CAREER PATHS

5

Introduction to the Mathematics Career Paths

Now that you have considered the scope and sequence of a successful job search, you are ready to survey the four major mathematics career paths and the variety of jobs within each one. Becoming involved in this kind of organized overview is interesting and fun for a math major, as it deals directly with the professional world you have prepared to enter, and it also provides concrete information with which you can visualize a future role in that world.

Before we begin the process of outlining each of the four main mathematics career paths, we want to offer one word of advice based on the important distinction between *contextual* skills and *portable* skills. If you are to be successful in setting and reaching your goals in a focused career path, you need to bear in mind the difference between these two types of skills. As you move forward, each job will provide the opportunity to develop skills in both of these areas. Both are important and both may be enjoyable, but you will need to remember that it is the portable skills that will help you most in your career.

Contextual Skills

These skills are specific to the industry in which you are working. For example, if you are an operations research analyst working in aircraft production, you will automatically learn a tremendous amount about how aircraft are built. That information is part of the *context* of your employment. It is important to recognize both the value and the limitations of that knowledge.

For example, if you change jobs and move to a job for a manufacturer of small motor equipment (snowblowers, chain saws, lawnmowers, and so forth), you cannot do much with the contextual knowledge about aircraft. You can, however, take your portable skills with you to your new employer.

Portable Skills

These skills include your *critical thinking* skills and your ability to *diagnose problems*, such as dealing with the question "What's wrong with this assembly line production setup?" Your ability to conduct a *needs assessment* is a portable skill: "Tell me about what happens when the production line goes down and what you have to do to bring it back up." Your abilities to *listen, design new systems, test prototypes,* and *build feedback systems* are all examples of portable skills. You will be able to take these with you to any new job that you may have in the future.

Your portable skills add significantly to your personal skills bank. Take advantage of all the opportunities your employer gives you to learn portable skills. If computer training is offered, take it. If there's a chance to learn a new job through cross training, grab it.

As you read the chapters that follow, think about your résumé. Imagine taking one of the many jobs described and working there for a year or two. How would your résumé change? What new skills would you be able to document? How would a new employer be able to evaluate your success in your previous job by the growth and accomplishment that you would be able to describe?

Chapters 6, 7, 8, and 9 present realistic job options for candidates with a mathematics undergraduate degree. The authors have designed, investigated, and written each of these chapters based on both the current job market for math majors and the knowledge we have gained in our own experience in counseling and advising students over the past three decades. Here we provide a brief overview of each of these chapters.

Chapter 6: The Math Job You Know Best

Your mathematics teachers have obviously played an important role in stimulating your own interest and enthusiasm for math over the years, and many probably conveyed some of their excitement about the possibilities and rewards of a teaching career. If you love your math studies, at some time during your college career you are going to ask yourself, "Could I teach mathematics?"

If you could talk with some of your teachers, you would find that they would tell you that good teaching, of math or any other subject, is about communicating to your students. We have all had teachers who knew their subject well but somehow could not get it across. It is "getting it across" that is the greatest part of the challenge in this career.

One of the most important ingredients for a teacher's success is that of genuine *appreciation* of the students and understanding where they are in their development and what they need at the present time. *Creativity* is also important if you are to engage your students' imagination and sense of fun and adventure as they explore new math concepts.

Planning is also a critical skill. If students are to succeed after they leave your class, you must accomplish certain goals. If you are teaching Algebra I, you must cover everything that the students will need to know before they enroll in Algebra II. That one great task takes many hours of planning and mapping out your lesson plan strategy to get from September to June without leaving anything out.

Never allow yourself to think that a teaching career is static. When you read Chapter 6, you will see that a career in **teaching mathematics** is vital, ever changing, and newsworthy. Our country is currently very concerned with the sharp drop-off in math and science interest, especially among young women, during the junior high school years. Exciting new initiatives are being implemented in curriculum, school design, and teaching interventions to correct this problem, and a great deal of work is still left to do. As a mathematics teacher, you could be part of this exciting change.

You should also be aware that recent IEA International Mathematics and Science Studies tests administered to high school seniors around the world have shown U.S. students falling far short of European and other industrialized nations' students in statistics, physics, and principles of science.

The results of these studies have had educators and the government calling for reenergizing our math and science teaching curricula. As career counselors, we are also seeing strong and continuing demand for math and science graduates. If you are interested in being part of the preparation of today's youth for math and science careers, this is a wonderful time to enter the teaching field.

Chapter 7: Jobs Using Math as the Primary Skill

Chapter 7 examines the four traditional career paths that require math as the primary job skill. All four of these paths and the many jobs within each path can use every bit of the mathematics knowledge and skill that you have acquired, and they will also challenge you by demanding more and more skill development and knowledge as you go along. If that is an exciting prospect, then give this chapter a close reading

The careers of **actuary**, **mathematician**, **statistician**, and **operations research analyst** all function in different areas of the economy and in very different job sites, but they are clustered in Chapter 7 for several specific reasons. Each of these career paths demands specific personal attributes and skills. Over and over, you will see that each path emphasizes working with teams, collaboration, and a need for the ability to explain and teach things to those who do not have a math background or vocabulary. In fact, the requirement of *superb communication skills* surfaces again and again.

People do not always associate being a skilled communicator with being an accomplished mathematician or scientist. Think of the physicist Dr. Stephen Hawking and the scientist and mathematician Dr. Carl Sagan, both superb communicators who have done outstanding jobs at helping us understand the mysteries of math and science. The quest of Dr. Alexander Wiles of Princeton University to solve Pierre de Fermat's notoriously difficult formula has been artfully communicated to millions. The story has actually become popular reading and TV fare for many people who would not ordinarily even approach the subject of prime numbers or, if they did, would not expect to understand it in the way that these books and films have revealed the subject for us.

Mathematicians must be strong communicators for the jobs described in Chapter 7. Actuaries, mathematicians, statisticians, and operations research analysts are frequently part of *management teams*. These teams form and re-form constantly, but the mathematicians are often a constant; they are called on frequently and usually to help advise and determine organization policy and direction, based on their projections and estimates of future behavior as predicted by the models they have established. Often, they are the only members of the management team who understand the math involved.

To make your points clearly and persuade effectively, you need to be very skilled at explaining complex information in simple language. Even though senior management positions for you are likely to be some years down the road, potential employers will still be looking at your résumé, cover letter, and interviewing style for evidence of your communication skills, even as you apply for entry-level positions.

Chapter 8: Working Toward an Advanced Degree

Sooner or later, someone will ask you if you have plans for an advanced degree. Even if you have selected mathematics education as your major in anticipation of teaching in public schools, you will find going on for an advanced degree is common there as well.

As you know, mathematics is a huge body of knowledge in which new discoveries are continually being made. Because there will always be more to learn, you may feel that you need to add to your math study after obtaining your bachelor's degree. If they can afford it, many undergraduates who are contemplating teaching positions in higher education will often continue with their formal education straight on through graduate school. Others, especially those interested in using their math in business or government jobs, are not as sure about an advanced degree and would rather get a job and make the graduate school decision later.

In Chapter 8, we advise thinking strategically about the kinds of jobs you might take after graduation that will do the best for you in terms of any future education. For example, the jobs we suggest as **marketing**, **research**, or **financial analyst** (the analyst jobs go by many different titles) will use your math skills to a high degree and provide you opportunity for impressive professional development. These jobs bring the chance to work with colleagues of a high level of managerial skill and educational background. More important, you can have the opportunity to use your undergraduate degree in math and earn an excellent income, while at the same time positioning yourself for a possible return to graduate school.

"Stepping out" of the academic world for a while before graduate school allows you to take a breather, use your degree, meet people in the field, and perhaps more realistically assess what you want in the long run. At the same time, choosing a first job as a research or financial analyst keeps your skills fresh and your income high. It also provides you with employee benefits for graduate education should you choose to return to school part-time. If you are hesitant about graduate school, this path gives you flexibility and options. Of course, if you have a once-in-a-lifetime opportunity for a grant or scholarship, or an extraordinarily favorable combination of professors and individual study opportunities in a graduate school situation, these factors must also be weighed in your decisions.

Chapter 9: Math in the Marketplace

The options presented in this final chapter are provided in recognition that some math majors have chosen their degree concentration simply because they were good at it and wanted a major that would allow them to succeed. Math may be something that you do well—even very well—but it may not be all that you want to do. Or, you may really enjoy mathematics and be very happy you majored in it, but perhaps your grades don't perfectly reflect the height of that enthusiasm. Whatever your motivation, Chapter 9 contains a number of job options for the mathematically talented.

Mathematics is important in the career paths that are discussed in this final chapter, but it is not the primary or even the most essential skill required. It is not necessarily more important than the ability to make a decision, solve problems, manage your time, or communicate well. The jobs in this path are done best by those who are comfortable with quantitative thinking and problem solving, but they may also be done by those whose strongest skills are in areas other than math.

Retail buyer, sales representative, and **purchasing agent** are sophisticated jobs, often responsible for decisions involving large sums of money. Interviewers and employers will be interested in your math skills but more concerned about your ability to think on your feet and be decisive and creative at solving problems. The jobs in this path can take you into any number of employment situations, including major department stores, federal government procurement offices, or any major industry or service organization from the airlines to financial institutions. The possibilities are endless.

Consider this final chapter carefully if you are interested in broadening your horizons with your math degree and looking at a variety of fields that offer not only opportunity to capitalize on your degree but enormous flexibility and potential for movement as you advance in your career. We guarantee that you will be pleasantly surprised at all these careers have to offer.

Many of us assume we already know all there is about careers in buying and selling. Like so many similarly unknowing assumptions, this one is usually wrong. Sales and marketing jobs, including the purchasing agent positions we describe, have become extremely sophisticated in modern times, requiring strong computer skills, a demand for detailed product knowledge, and far more listening and consultation than persuasion. These jobs are interesting, challenging, and also worthy of your attention.

Additional Resources

Suggestions for additional resources are found throughout the book. These publications and organizations offer a vast amount of information in a variety of media and points of view, and we hope they will be useful.

6

The Math Job You Know Best

Mathematics teachers have already influenced your student life in significant ways, and you have observed them throughout all of your student life. Your concepts of effective teaching practices and goals and your thoughts and beliefs about the value of teaching have developed over many years. Now you are considering a teaching career of your own, and a new aspect of this career has suddenly become very important.

You may never have thought about teaching as a job or a career and may not have considered the ways in which teaching is a job like any other job. Salaries, benefits, unions, bosses, and coworkers are all part of teaching, and the most important practical factor of all is the job market itself.

If you are considering a career in teaching, you must have a realistic perspective on the job market and its projected future. Jobs in the teaching field, just as those in many other fields, have been affected in recent years by the overall condition of the nation's economy. In some markets, it is difficult to break into teaching. In 2005, the federal budget eliminated or cut back funding for certain educational programs, and additional teaching positions disappeared. But the good news for mathematics teachers is that in math and science there is a distinct shortage of teachers in many places and jobs are available.

Perhaps you may not immediately be able to get a job in your first choice of location, but there are jobs in many areas. Because you have specialized in mathematics, your chances are better than average among job seekers in general. According to the U.S. Department of Labor, the projected job growth for mathematics teachers is expected to stay above average through the year 2012.

The Teaching Challenge

Teaching mathematics requires much more than just knowing mathematics. Even though the subject is mathematics, the most important concern in teaching it is not just having the skills and information but being able to instill appreciation and enjoyment of math in all its myriad forms. This concern must be paramount, because the teacher's responsibility begins not with the subject area but with the student. Learners come to a math classroom with different issues. They come at different ages, for different reasons, from different lifestyles, and with dramatically different degrees of interest—and anxiety—about math and the teacher who will teach them.

If you talk with math teachers, they can probably assure you that they don't teach *math* as much as they teach *students*. The art of teaching and the skill required in the dynamics of teacher-student interaction weigh much more heavily in the learning equation for teachers than their own love of the subject matter.

You will quickly recognize that in a teaching career, simply having a love of math is not enough, although that is certainly important and desirable. How could you begin to teach something you did not truly enjoy without conveying your disinterest through a dry and mechanical approach to the subject? However, teaching any subject requires many additional and different skills than those demanded by studying the particular discipline. We have all observed that the world is full of skillful practitioners who, for one reason or another and quite often inexplicably, cannot teach someone else how to do what they do.

Teaching skills and aptitudes include instructional methods, management, empathy, communication, organization, and many others. To begin with, *planning for learning outcomes* is critical. Teaching math within an established curriculum—in middle school, high school, or college—means corresponding to some stated goals or course outlines. In middle or high school, the curriculum must adhere to strict and detailed state standards, as well as standards defined by the local school district. In college, it may be defined in a written course description created cooperatively by the math department and the college administration and carried out in the syllabus that you yourself prepare for each course.

Meeting Learning Outcome Goals

Accomplishing a specific body of learning within a set time period requires judicious planning. What will be done each day? How much time should be allowed between assignments or readings? Which materials must be required and which should simply be recommended? In every teaching job, you must make countless decisions. Which textbooks and ancillary materials will you use, and how will you evaluate your students?

Student bodies today are very diverse. Your students will come from a wide variety of backgrounds in different educational systems. They may learn better in different modalities. Some are auditory learners who enjoy listening and concentrated listening is how they best absorb material in the classroom. If they are required to take notes while they listen, it may be difficult for some of them to retain the material. Other students prefer a visual approach with board work, handouts, models, videos, their own notes, diagrams, books, and other visual materials. They retain images best and can call them up to remember the substance of the material.

Other students need to participate through activity and physical involvement with the material. These are kinesthetic learners, and they are often forgotten in course planning and curriculum design. These learners like to participate in team projects or contests such as building a pyramid or cube, going outside and estimating the height of a building, and participating in other activities that physically involve them. Those who learn best this way must also be factored into a teacher's planning.

A successful teacher ensures that the learning styles of all the students are satisfied through presenting a judicious combination of modalities in the students' learning situations. Good teachers must analyze their own teaching styles and work to make certain that they are incorporating the elements that may come less naturally to them but may be essential to some of their students.

Dynamic and Sequential Learning

The teaching and learning that take place in the classroom are not static. The classroom is an emotionally charged environment for the student and instructor that may call into play questions of self-esteem and competency. Students are continually exploring themselves in relation to their capabilities, values, and achievement. A good teacher understands this and encourages a risk-free environment of mutual appreciation and participation. Both teacher and student must be allowed to make reasonable mistakes and then move on.

The teacher strives to assist in establishing congruence between the real self (how we think of ourselves at that moment), the ideal self (who we want to be), and the learning environment created in the classroom. It is hoped that the classroom will be a place where the real self can rise up and begin to become the ideal self.

While any classroom can cause us to call into question who we are or how competent we are, this is perhaps especially true for the mathematics classroom. Math requires an orderly sequence of skill development, like a graduated set of building blocks. Certain skills must be mastered to move on to other more advanced proficiencies. Many students develop a strong sense of inadequacy early on and their own levels of competence. You've heard this from your friends and family, and you'll continue to hear people's "horror stories" about math every time you mention that you are a math major. Good math teachers can do a great deal to alleviate this sense of inadequacy by developing methods that will help their students get past it and move ahead to a sense of confident growth.

Meeting Resistance to Learning Mathematics

In many—perhaps even in most—mathematics classrooms, there will initially be a degree of resistance to learning math. The mathematics teacher must be prepared to establish a healthy attitude in the students and help them develop confidence in trying new tasks and finding satisfaction in their progress.

Traditionally, resistance to mathematics and science education has been particularly detrimental to young women. For many years, researchers have been interested in the educational and career barriers for females in the field of mathematics. Studies (e.g., Fennema and Sherman, 1977; Fox, 1980; Armstrong, 1979; Boswell and Katz, 1980) have shown that the development of attitudes that can affect a female's math achievement begins at about thirteen years of age. It has been established that these attitudes are a result of stereotyping, not a result of aptitude. These studies make clear that, at this age, many talented young girls are rewarded for social conformities of various kinds and many may move away from math for these reasons.

Fortunately, studies have alerted teachers, other educators, parents, and employers to the problem, and many have been working to encourage young women to enter fields that center on math and science. The increase in women's employment in mathematics-related careers has demonstrated the importance of this change in attitudes, expectations, and encouragement.

Evaluating Students' Work

Grading and evaluating learning is another important area of concern to the math teacher. Grades are an expected and required part of classroom culture in many institutional academic settings. Establishing fair and consistent standards of evaluating students and assigning grades is sometimes a significant challenge for many teachers who otherwise feel perfectly competent in their teaching roles.

Math teachers find that having regular assignments is helpful in monitoring students' progress. Daily homework and frequent quizzes and tests may help in tracking a student's progress and provide feedback for the student.

But these activities result in a lot of work for the teacher, and much has to be done outside the classroom, during office hours or at home in the evening. So, if you're looking for a nine-to-five job, teaching is not for you.

Motivation and Application

Many other roles are also required of the math teacher. Animating the class and inspiring attention and commitment to the material are all required in teaching. Part of this is accomplished through the teacher's enthusiasm, teaching style, effective use of ancillary materials, and ability to relate the material to the students' lives. Of course, math teachers present information, but they also give students lots of examples of practical applications.

Math can become especially interesting when young people understand how they can use it in their daily lives and how the knowledge and skills can make their lives easier and more productive and satisfying. It takes creativity and energy to bring together math and its applications to the real world in the classroom and demonstrate math's tremendous value and usefulness to the students.

Additional Teacher Responsibilities

Raising relevant questions, prompting dialogues within the class, and developing the discipline of self-questioning within the students are all important additional responsibilities of the math teacher. Teachers also clarify students' confusion and lead them out of difficulties by drawing parallels and finding relationships between examples.

As well as teaching math, a math teacher shows students how to learn. The skills involved in learning to question; retaining information; being selective; recording, ordering, and organizing information; and solving problems are priorities of each day of effective math teaching.

It is essential for a teacher to use the class and the material to model good work and study habits. When teachers share their personal enthusiasm for mathematical ideas in discussions of the material under study, students will be more highly motivated and interested in the work as well.

An instructor will also, by example, develop the students' capacity for self-evaluation through careful, caring feedback about the students' work. The instructor's own example of preparation, organization, evaluation standards, personal appearance, interest in students, and enthusiasm for the subject may remain an example long after the memory of an actual class's content has faded.

Teachers are often cited as having been exceptionally important factors in their students' choices for their careers. Very often, too, teachers themselves will remember one or more of their own teachers who were strong influences on their decision. Much of that influence is most likely a result of a teacher's classroom persona. A special teacher serves as a model of someone who enjoys what he or she is doing and who enjoys doing it skillfully. Such special teachers are professional and correct, yet remain natural and approachable. We can observe them as they work each day in the classroom and be encouraged to think, "I would like to do that work, too."

Definition of the Career Path

In this chapter, we will look at two of the possible levels of teaching math: secondary school teaching, with a bachelor's degree, and college teaching, possibly with a master's degree, but more frequently with a doctoral degree as the essential credential.

Teaching in Secondary Schools

Licenses and certification are required in all fifty states. Most of the states require an education major as well as a subject area concentration for secondary school teaching. If you are interested in teaching at the middle, junior high, or high school level, you'll need to major in math education so that in addition to your math courses, you acquire the necessary education courses to meet state certification competencies. In some schools, you will be required to complete all the courses of a full mathematics major and, in addition, take the required education courses. In others, a mathematics education major may require fewer advanced math course requirements and be a balance of mathematics and education. In either case, the requirements for teacher certification in that state are usually built into the curriculum, and you will be prepared for examination for certification by the time you graduate.

All certified teacher graduates are qualified for entry-level math teaching assignments. In some situations, the first-year teacher's inexperience can actually be considered a plus. With school budgets currently under strain, principals, superintendents, boards of education, and other hiring officials may be more attracted to a new teacher who will earn a lower salary than someone more experienced with a higher degree who would require a larger salary.

Most states require continuing education, and teachers must take a certain number of courses to renew their licenses. A typical requirement is the acquisition of an additional six credit hours or continuing education units of graduate work for each renewal of the license or certificate. Continuing education courses can sometimes be obtained in distance-learning classes, e-learning classes via the Internet, or professional workshops or seminars offered by professional education organizations or by the school districts themselves. Some systems also require teachers eventually to get a master of education degree, and some school systems pay all or part of the tuition for these courses.

A frequently asked question is "Is it possible to teach math at the high school level without state certification and with a bachelor of arts in math?" The answer is yes, in some circumstances. In some public school districts that have had difficulty securing teachers because of location or pay scales, provisions have been made to grant temporary certification to noncredentialed teachers. This situation often occurs in school districts that for reasons of geography or economic circumstances, do not attract enough applicants. Some private high schools may consider a noncertified teacher. A school district may sometimes also need a long-term substitute position for a secondary math teacher because of an illness or some other leave of absence. Depending on the pool of applicants, a noncertified teacher may be able to secure such a long-term substitute position.

Teaching fellowships, residencies, and internships are offered by some school districts and often involve intensive summer-training programs before the school year begins, plus the opportunity to become certified and in some cases to continue study for a master's degree. Some school districts may be able to use fellowships or internships to fill vacancies that they might not otherwise have been able to afford to fill in a given school year. In other instances, a fellowship, residency, or internship program may be possible because of government grants and funding sources. A good source of information for internships and fellowships can be found on the website math-jobs.com, which also gives links to other sources. Following is a sample ad for a noncertified teacher:

Math Teacher: Outstanding professionals and recent college graduates; no education experience or courses required; will teach middle school and high school students in this city's highly challenging Teaching Residency Program; intensive summer-training institute; residents will begin to teach in fall semester while taking additional classes to obtain teaching certificate. Will teach full-time for regular full-time salary and benefits. U.S. Citizenship or Permanent Residency required. GPA of at least 3.0. Residents will receive up to 75% subsidized tuition and opportunity to pursue master's degree.

All noncertified teachers who secure provisional situations or long-term substitute jobs will eventually have to be state certified. Although many states have reciprocity agreements, teachers often have to get additional training to qualify for certification in a new state. Private schools increasingly require teaching credentials that equal those of public school's. At many private schools, it is not uncommon for math teachers to have master's degrees. Numerous large-city high schools have attracted teachers with advanced degrees, including Ph.D.s.

Teaching with a Master's Degree

A master's degree in math may be helpful in securing a private school teaching position at the high school level, especially if the master's work in math corresponds to the school's needs. Advanced degree work in a master's degree program frequently provides the opportunity to assist a faculty member in teaching an undergraduate class and may include experience designing exams and grading tests, as well as staffing math-help clinics.

Graduates with master's degrees and no certification at the bachelor's level may also find employment in junior and community college settings or special college programs for adult learners. These schools may welcome the teacher with a master's degree, especially if the specialty is one that will be useful in their curriculum. The following sample posting for a master's degree candidate working as an instructor of mathematics in a two-year college includes a wide range of duties:

Mathematics Instructor: Essential functions include teaching assignments, which may range from developmental mathematics through differential equations. Master's degree in mathematics required. Preferential consideration given to candidates with at least two years of teaching experience and knowledge of graphing technology and computer assisted software (CAS).

Teaching with a Doctoral Degree

A doctoral degree in mathematics opens up the world of college teaching to the prospective educator. There is a high degree of competition for these positions, but at the time of publication of this volume the number of college teaching positions and the number of new Ph.D.s in mathematics indicate better-than-average possibilities of securing employment.

Most college and university websites provide detailed information on jobs and careers at the school or at least will provide a list of openings with contact information where you can find additional details.

Positions are also advertised in periodicals such as *The Chronicle of Higher Education*, which reports on higher education and contains a comprehensive listing of faculty, staff, and leadership position openings for colleges and universities in the United States and some foreign countries. The following is an ad from *The Chronicle* that would be of interest to a new Ph.D. in math who has some experience teaching at the undergraduate level:

Mathematics: Four-year liberal arts college is seeking applications for a tenure-track position. Candidates should have a Ph.D. in mathematics (preferred) with expertise in applied mathematics and/or statistics and strong evidence of excellence in undergraduate teaching. The successful candidate will be capable of assuming a leadership position within the department in the future; will have experience in using current math pedagogy including technology; will be expected to teach all levels of undergraduate mathematics; will participate in student advising. Salary relative to experience.

This ad, interesting for a number of reasons, requires first of all an earned doctorate. To apply, you must have your degree completed and in hand. It is a position that has been reserved for tenure line, so scrutiny of candidates will be intense—all the more because the ad indicates the candidate must display the potential for assuming a leadership role, probably as department chair, at some point in the future. Nevertheless, the undergraduate course load stated suggests that the position is open to an exemplary new doctorate who may have accumulated significant teaching experience previously in another institution or while teaching in graduate school.

Some ads will encourage the application of ABD (all but dissertation) candidates who have completed all required doctoral course work but have not yet completed the writing of the doctoral dissertation. A position for a candidate at the ABD level will generally not pay as well as a position for a candidate with an earned doctorate and will not lead as directly or as quickly to possible tenure and promotion. ABD candidates will also have to decide how they will finish their degree—the dissertation often being the most time-consuming aspect of their academics—while they are holding down a full-time job.

Introductory college math courses are usually part of the teaching load of new college math teachers. Many of your students will be taking some variation of an introduction to college math course or whatever the institution requires for students to meet the college's general education requirements. Students enrolled in these courses are taking them because it is a college requirement for graduation and not because they are math majors or have chosen the course for personal interest. The math department performs a general education service to the entire college in offering this course. Usually, even senior faculty members will teach at least one class of first-year college math, although, as you become more senior in the faculty, you can add courses more directly related to your specific interests and educational background.

Remember that despite your advanced degree and specialized work in mathematics, you are very apt to have a teaching load that is composed mainly of lower-level courses. You will undoubtedly have opportunities to pursue your doctoral field in writing and research and speaking at scholarly conferences, but you may have to teach for a few years before your teaching schedule includes advanced students.

In the previous ad, candidates applying for this position must be able to document their prior college teaching experience. This could come from teaching assistantships done while working on their doctoral degree. Many students acquire this experience as graduate teaching assistants, part-time faculty, lecturers, or adjunct faculty at other colleges.

The road to a doctorate is long and arduous. Along the way, you will meet some wonderful people, some of whom will be friends and colleagues the rest of your life. Even colleagues separated by long distances have the opportunity to revisit at conferences and symposia. You will also have opportunities to write, teach, and perhaps publish—all before you finish your doctorate. Take advantage of these opportunities when you can. It is possible, however, to become overly involved in some of these activities to the detriment of your degree progress.

It has been a concern in academic circles that a significant number of individuals begin doctoral programs and do not complete them. As president of Harvard University, Neil Rudenstein published a book on improving and tightening up the time requirements to earn a Ph.D., particularly in the humanities, where his research demonstrated the longest timelines between initiating the degree and earning it, with a correspondingly high rate of dropouts. In science and math areas, he found a higher completion rate and shorter time-to-degree completion. This news is encouraging to the potential Ph.D. candidate in math.

Working Conditions: High School Level

Working conditions for teachers of mathematics can vary dramatically, according to the particular educational setting. The high school math teacher has a full complement of classes, perhaps as many as five or six a day, and may have study hall or lunchroom supervision duties during the week, responsibilities for some after-school study clubs or detention centers, or even a sports activity to supervise.

Perhaps the single most challenging element of working conditions for any secondary school teacher is that of discipline in the classroom. Having been a high school student yourself, you realize that attendance is for the most part not voluntary so students sometimes exhibit resistance and acting out is not uncommon.

Effective classroom teachers must successfully master classroom management. For many young teachers, this is the most challenging part of teaching and may also make for the most interesting stories as they grow in their profession. The time spent teaching math and maintaining classroom discipline is seldom in balance, and it can be frustrating when one disruptive student threatens the decorum or even the safety of the entire class.

Most public high schools have fairly rigid rules of behavior for students, and the primary agents of enforcement are the faculty. To elect high school math education as your particular arena is to challenge your ability to maintain your poise and focus on the subject matter while at the same time enforcing and administering the necessary discipline mandated by your school. This includes grading students' work; making referrals to counselors and the principal; assigning detention; issuing warnings; attending conferences with parents; and cooperating with police, social workers, and youth workers if needed.

The secondary teaching day is full, with clear starting and ending times, plus much at-home work for every teacher. Lesson planning is also time consuming, as is maintaining the required records of attendance, grades, warnings, progress reports, and other evaluations that may be required in your school district. You may also frequently find yourself speaking with parents by telephone from your home in the evenings.

The workday doesn't stop when the final classes are over. High school teachers are required to accompany students on field trips, be speakers, chaperone dances, or advise the yearbook or literary journal student staffs or other clubs in the school. These duties can be time intensive, and it is important that the teacher entering into secondary teaching understands that these assignments are not so much additions but typically an integral part of what makes up a high school teacher's commitment.

Working Conditions: College Level

The college teaching environment is significantly different from a middle or high school setting. Classes, office hours, meetings, and research work are all part of the schedule. Because college campuses are often centers of art, music, and intellectual exchange, there are also frequent events to attend in the evening. Faculty members may act as advisers to fraternities and sororities, campus newspapers, or other activities and clubs, which will add more commitments to their schedules.

The college day is certainly less rigid than that of the high school's, though it may be just as busy and long. The difference is that, for high school teachers, the day is largely predetermined. The college teacher, on the other hand, may feel institutional and professional pressures to fulfill certain roles, but the actual determination of how to do most of that is left up to the individual. There is less need to appease a number of outside publics. There are no principals, school boards, parents, parent-teacher groups, or civic groups to satisfy.

The college classroom is closed to outsiders and is not violated by anyone outside the class. This convention is so well understood that it is rare to have a class interrupted by anyone from outside of the room. Academic freedom protects professors in large part and allows them to express themselves within their class material with far greater freedom than is the case in high schools.

Methods of grading, evaluation procedures, number of tests, and even the issue of whether to have textbooks are almost entirely up to the faculty member, and if a good rationale supports these decisions, the college usually will not interfere. An added protection is the granting of tenure to established professors who have documented significant teaching histories and excellent student reviews, publications, campus committee work, and outreach to the community. With tenure, faculty members are secure in their jobs for the rest of their professional lives, barring any unusual event or misconduct that could cause them to be released.

Granting tenure to established professors gains them an additional degree of job security and further supports their expression of academic freedom. With the financial pressures of recent years, the tenure system has come under increased scrutiny. There are movements to initiate measures to determine faculty productivity, but in most schools, these are in the earliest stages of exploration and development.

It is important to note that, in spite of these apparent freedoms, standards of accountability still apply, and some of these standards are not very different from those of the high school teacher who must meet state educational requirements. If you are teaching a basic college course in algebra, it will become apparent to your colleagues in your department if you are not covering the material adequately. When they receive your students in upper-level classes, they will expect you to have covered certain material and to have taught it well. So, while college teaching brings freedom of choice in methods and pedagogy, it in no way frees you from accountability for accomplishing your teaching task.

The actual teaching day in a college or university setting involves fewer class hours taught per day and per week than in a high school setting. At an institution that focuses on faculty research, the teacher would be responsible for teaching two or three courses that meet approximately two hours per week each. Colleges that emphasize teaching rather than research require instructors to teach three or four courses, for a total of eight to twelve hours of class meetings per week. These class hours and some mandated office hours for advising students and other general advisees are the principal requirements for attendance on the faculty members' part. Additional time for preparation, grading, and personal study is to be expected. Following is a sample ad for an assistant professor:

Mathematics: Tenure track position in mathematics at the rank of assistant professor. Ph.D. in mathematical sciences required. Strong commitment to teaching required. Proven teaching ability with experience in technology and a background in statistics is preferred. Responsibilities involve teaching a wide range of undergraduate mathematics and statistics courses; normal teaching load is twelve hours/semester. The successful candidate will also be expected to participate in academic advising, faculty committees, and professional development activities and related community service. Salary is dependent upon educational preparation and experience.

In addition to teaching courses and advising, scholarly research is expected even at those colleges where tenure is not based on publication. All colleges want their faculty members to contribute to the scholarly dialogue in their disciplines, and this production is reviewed by chairs of departments and academic deans periodically throughout the instructors' careers. Publication may be a determining element in granting tenure or promotion. It may also influence issues such as salary negotiations and merit increases. An old maxim says, "Publish or perish!" While this expression may sound like a joke, and it may actually—partly—be one, it also has truth in it. A college teacher must recognize the role of published work in a successful career.

At most colleges, the faculty also comprises the governing and rule-making bodies that determine and vote on governance and program changes. Committee work can be issue oriented, such as a commission on the status of women or a faculty pay-equity survey. It may be programmatic, such as a committee to study the core curricula for undergraduates or to devise a new graphic arts major. Or the work of the committee may be related to credentials, as in a committee set up to prepare materials for an accreditation visit by a state agency.

In most colleges and universities, some committees, such as those on academic standards, curriculum review, promotion and tenure, planning, and administrator review, are permanent, although the membership may change on a rotating basis. Other committees are formed for a limited time or until the completion of a particular task. These committees are essential and are vehicles for guiding the direction of the college. Having the support of all the faculty and constantly fresh and, therefore, more interested members helps to ensure that all voices will be heard and that many different opinions are considered in making often far-reaching decisions.

Training and Qualifications

Requirements from state to state are fairly uniform. To teach math at the secondary level requires a bachelor of science degree in secondary school mathematics education and state certification for the state in which you wish to teach.

These secondary education programs are well-defined options within the education curriculum of many teacher training colleges and universities. They include student teaching, where you have the opportunity to leave campus and teach actual math classes under a supervising teacher, usually for one semester.

Certification in the state granting the degree is usually part of the degree process and may include the requirement to participate by taking a national examination. You will hear reference to the National Teacher Examination (NTE) frequently. NTE, however, has become a generalized term that often refers not just to the original NTE (a specific exam) but to other national professional assessment instruments such as The Praxis Series.

You should always consult the state department of education for information about the tests needed in a particular state. Current state-by-state information is available on the Praxis Series website at ets.org/praxis. The Praxis Series is used by most states that require testing and has three categories of assessments: (1) Entering a Teaching Training Program, Academic Skills Assessments; (2) Licensure for Entering the Profession, Subject Assessments; and (3) First Year of Teaching, Classroom Performance Assessments. In nearly all states, test preparation is now well incorporated into the college curriculum.

Master's Conversion Program

If you already have a degree in mathematics and you want to teach at the middle school or high school level but do not have a certificate, another option is to enroll in a conversion program.

These programs provide the opportunity to add the necessary state-mandated teaching requirements to your existing degree. Completing the certification requirements could take from twelve to eighteen months of academic enrollment or, in some cases, a full two years. The amount of time depends on the courses you have taken in preparation for your undergraduate degree and whether a change of institution is involved.

In some cases a conversion program can provide you with the necessary certification requirements and also confer a master's degree. A master's conversion program is attractive but can sometimes present difficulties in a new teacher's job search, depending on the teacher-candidate supply and demand in the job market. Because your school district must pay you more if you have an advanced degree, the principal may find you somewhat less attractive than a comparable candidate with a bachelor's degree in math education and similarly limited experience. If math teachers are in short supply, however, those hiring may be delighted to find you and will pay the salary increase that your new degree requires.

Some conversion programs may also be available independent of a collegiate institution. For example, some are offered as teacher training institutes by a consortium of school districts. These teacher qualifying programs may also be offered by associations, unions, and other involved organizations. The training institute will accept those with bachelor's degrees, many of whom have had other careers or significant work experience, and place them with master teachers in actual classrooms for a full year. Half the year may be spent with one grade and the remaining half of the year with another.

The year of classroom experience includes much independent work and follows a contract established at the start of the year. The contract specifies how and when the student will "solo" in the classroom, although, in practice, these partnerships between supervising teachers and interns usually allow for the intern to acquire significant independent teaching mastery while working with the supervising teacher. Specific learning outcome goals are set to ensure that the program delivers the requisite training and experience for certification. In some places, teachers may also be required to participate in an associated classroom training program to meet state certification requirements.

Teaching in Colleges

The candidate for a college or university teaching position must have a doctoral degree or, in some cases, all but the dissertation, or ABD. If the candidate begins teaching as an ABD, salary is less and assignments are fewer than with a Ph.D. In addition to a doctorate, as we have seen, the candidate may be required to have done a certain amount of teaching, research and publishing, or practice in a particular genre or subject area in math, plus some additional competencies. The hiring institution almost always requires experience in teaching introductory college math classes to first- and second-year undergraduates, which most college teaching job candidates will have done as teaching assistants (TAs, sometimes also called graduate assistants, or GAs) during their master's or doctoral work.

Earnings: Middle and Secondary Schools

Math teachers at this level are paid according to the same salary schedules as other teachers in their school districts, and these schedules are based on a combination of degree level and teaching experience. Some school districts pay a small additional amount for each graduate school course credit if the candidate does not yet have a master's degree. These salary schedules are public information, and you can obtain them from the state's department of education or from the local school district.

Teachers' salaries vary dramatically across the United States by tens of thousands of dollars, depending on the affluence of the school district, the

BEGINNING TEACHER SALARIES BY STATE, 2001–2002

State	Salary	State	Salary
Alabama	$29,938	Missouri	$27,554
Alaska	36,294	Montana	22,344
Arizona	27,648	Nebraska	26,010
Arkansas	27,565	Nevada	28,734
California	34,180	New Hampshire	25,611
Colorado	28,001	New Jersey	35,311
Connecticut	34,551	New Mexico	27,579
Delaware	32,868	New York	34,577
District of Columbia	31,982	North Carolina	29,359
Florida	30,096	North Dakota	20,988
Georgia	32,283	Ohio	29,953
Hawaii	31,340	Oklahoma	27,547
Idaho	25,316	Oregon	31,026
Illinois	31,761	Pennsylvania	31,866
Indiana	28,440	Rhode Island	30,272
Iowa	27,553	South Carolina	27,268
Kansas	26,596	South Dakota	23,938
Kentucky	26,813	Tennessee	28,857
Louisiana	28,229	Texas	30,938
Maine	24,054	Utah	26,806
Maryland	31,828	Vermont	25,229
Massachusetts	32,746	Virginia	31,238
Michigan	32,649	Washington	28,348
Minnesota	29,998	Wisconsin	27,397
Mississippi	24,567	Wyoming	26,773

particular contract the teachers have in that district, and other variables. Because of this disparity, it is important for you to research the salary levels in an area carefully, being sure to get up-to-date information.

During the 1990s and on into 2001, the United States experienced an increasing teacher shortage in almost all fields. By 2005, because of declining classroom enrollments and other issues, that shortage had disappeared in many areas. In mathematics, however, the shortage still exists, so the job market for math teachers is better than for teachers in some other subject areas.

Beginning salaries vary from state to state. In general, the richest and most populous states and those with the highest costs of living pay the largest salaries. The average annual teacher salary for all teachers in 2001–2002 was reported as $44,714. Beginning salaries were somewhat lower. States paying the highest salaries were California, Connecticut, New Jersey, Michigan, New York, Pennsylvania, Massachusetts, Rhode Island, Illinois, and Alaska.

According to the National Education Association (NEA), the previous salaries represent beginning teacher salaries in each state for the years 2001–2002. For more information, check the NEA website at nea.org, where they provide information about individual state needs for teachers as well as information on how to contact each state board of education.

As of the 2004–2005 school year, nearly all fifty states reported teacher shortages in math, science, and special education. This does not mean that there were jobs available in all areas, because in some places a shortage existed but there were no funds available to hire more teachers. The American Association for Employment in Education has said that schools are facing the worst budget crunches they have experienced since the 1940s, and many school systems have had to lay off teachers and eliminate classes in art, music, social studies, and other areas.

The NEA lists mathematics as second only to multicategorical special education in teacher shortages across the country. Math teachers in most parts of the country, therefore, still have the luxury of choosing the school systems where they want to teach, and it is expected that they will be able to enjoy that luxury in the job market through 2012.

Earnings: Public and Private Schools

Public schools must abide by legal equirements for disclosure of salaries, benefits, and working conditions, and active teacher unions usually are able to negotiate for increases in salaries and better teaching conditions from time to time. Salaries in private schools are usually lower than those in public schools, and there may be no union or teacher organization at all. Other small perks and values, however, may compensate in part. For example, housing may be provided in an attractive rural setting, or free or reduced tuition may be available for the teachers' children. The teacher who is planning to work for a private school needs to be very thorough in researching the working conditions and the teaching contract.

Earnings: Postsecondary Schools

Postsecondary salaries are substantially higher than those for secondary schools, and the median salary for an associate professor of mathematics in

2005 was expected to be $57,481, with 25 percent making $88,724 or more, and the lowest 25 percent making $45,484 or less. Bonuses and benefits added from 10 percent to 15 percent additional value in most areas. Degrees; years of experience; course hours to be taught; and additional responsibilities such as leading faculty groups, research, publication, and other tasks had significant effects on salary levels.

Strategies for Finding Jobs

Your strategy for finding a mathematics teaching job will vary depending on your degree level, the grade you wish to teach, and the type of school where you plan to work. Different strategies will be required for finding jobs in a public or private school, or in higher education. There are some common strategies, however.

Public Schools

In many cases, the most direct approach is to send a cover letter and résumé to the schools that interest you. The state departments of education provide listings of all public schools in each state that give you the names of the superintendents, principals, personnel officers, or other administrative contacts. Not only do these listings give the names, but the telephone numbers and addresses are listed as well. Check with your career services office for these directories.

In other cases, you will find notices on the Internet stating that only electronic applications will be reviewed. In those cases, you will want to send your electronic résumé plus cover sheet and cover letter. In the cover letter, you can also offer to send a formal paper résumé if you wish.

Private Schools

Private schools seldom advertise positions in newspapers in order to have a more select pool of candidates and to maintain a lower public profile than their public school counterparts. Finding a job teaching math in private schools requires more research work and diligence than looking for teaching positions in public schools. You'll need to check with your library or career services office for a directory of private schools. These directories list important information about grade levels, tuition, number of students and their gender, faculty, special facilities, and a short history of the institution. Send a résumé and cover letter to each school where you would like to work.

Many private schools use teacher recruitment services for the discretion and efficiencies of time and money they can provide. Be sure to investigate the services provided and any fees required by you before signing on the dotted line. Also search on the Web at petersons.com/pschools for listings of current job openings in private schools and for specific instructions on how each of the schools advertising positions prefers that you apply.

Directories

Patterson's Elementary Education (Edition 2005, published by Thomson Peterson's) and *Patterson's American Education* (Edition 2005, published by Thomson Peterson's) contain listings of school districts, superintendents, and public and private schools. For more information on these two publications, consult the website at EDIUSA.com. Private schools can be found in *Peterson's Guide to Independent Secondary Schools*, the *Handbook of Private Schools*, or on the Web at petersons.com.

These books may be found in your career services office or a public or university library. If you are considering teaching at the college level, look at college guides to two- or four-year colleges such as *Peterson's Guide to Two Year Colleges*, *Peterson's Guide to Four Year Colleges*, and *Peterson's Guide to Graduate Study*.

Career Office Postings

College career offices post job openings that they receive directly from school districts, colleges, and other organizations, and they subscribe to various regional and national listings as well. Math teaching positions can be found in publications such as *Current Jobs for Graduates in Education*, *The Job Hunter*, *Current Jobs for Graduates*, and *The Chronicle of Higher Education*. These publications maintain Internet home pages where you can preview the listings free of charge or subscribe to them on the Internet.

Math Departments

Math departments are a rich source of information about job openings. Postings are sent directly to the departments, and the announcements are usually posted on bulletin boards or are available in a notebook binder in the department office. Talk with your adviser or the department secretary to find out what information is available and how you may have access to it. Be sure to check bulletin boards regularly to be aware of new postings as quickly as possible after they are posted.

Professional Associations

Professional mathematics and teachers associations provide a wealth of career and job search information. Virtually every one of them has its own website and publishes a variety of useful material. They often publish their own newsletters or journals where job openings are listed in a "Careers" or "Jobs" section. Review the list of professional associations at the end of this chapter. Check their websites to see whether they provide a job openings listing, and follow up on all the links and sources of additional information if you are interested.

Possible Job Titles

Most job titles will include the word *teacher*, and ads may list them as follows:

Cooperating teacher
Educator
Elementary school math teacher
High school math teacher
Intern teacher or teaching intern
Math/science teacher
Middle school teacher
Resident
Secondary mathematics teacher
Teacher

In higher education, the status of the position becomes part of the title, such as in the following:

Assistant professor
Associate professor
Instructor of mathematics
Professor of mathematics

In addition, teachers of mathematics can locate related positions in the following business areas:

Banking
Defense contract companies
Disease control and prevention centers
Engineering
Federal government
Health insurance companies

Human resources
Information science (technology)
Management and public relations
Mapping agencies
Research and testing
Science, engineering, and technical services
State government
Training and development

Related Occupations

Mathematics teachers share the ability to communicate information and ideas with other types of occupations. Here are some related occupations that require a degree in or knowledge of mathematics. Investigate these and other positions that use this skill base.

Consultant
Cryptologist
Employee development specialist
Librarian
Lobbyist
Policy analyst
Statistician
Trainer
Writer

Professional Associations

Following are some of the associations that are related to teaching mathematics. For more information about these professional associations, check the websites listed and also refer to the *Encyclopedia of Associations*, available at all major libraries. Review the Members/Purpose section for each organization, and decide whether the organization pertains to your interests. Membership in one or more of these organizations may be helpful in finding job listings, networking opportunities, and employment search services. Some provide career search information and job listings at no charge. If you want to receive the association's mail or e-mail publications, which provide ongoing lists of job opportunities, some may require you to join the organization.

Student member rates are usually available at a slightly reduced cost, and information about these and any other fees is usually published on the home website.

American Association of Colleges for Teacher Education
1307 New York Ave. NW, Suite 300
Washington, DC 20005-4701
aacte.org
Members/Purpose: Colleges and universities concerned with the preparation and development of professionals in education and human resources. Seeks to improve the quality of institutional programs of the education profession.
Journals/Publications: AACTE annual directory, AACTE briefs, *Journal of Teacher Education*

American Federation of Teachers
555 New Jersey Ave. NW
Washington, DC 20001
aft.org
Members/Purpose: Works with teachers and other educational employees at state and local levels in organizing, collective bargaining, research, educational issues, and public relations. Conducts research in areas such as educational reform, bilingual education, teacher certification, and evaluation.
Journals/Publications: *AFT Action: A Newsletter for AFT Leaders, American Educator, American Teacher, Healthwire*

American Mathematical Association of Two-Year Colleges
Southwest Tennessee Community College
5983 Macon Cove
Memphis, TN 38134
amatyc.org
Members/Purpose: Two-year college mathematics and computer science professors, and four-year college mathematics professors concerned with lower-division mathematics education. Encourages development of effective mathematics programs. Allows for the interchange of ideas on the improvement of mathematics education and mathematics-related experiences of students in two-year colleges or at the lower-division level.
Journals/Publications: *AMATYC News, AMATYC Review*

American Mathematical Society
201 Charles St.
Providence, RI 02904-2294
ams.org
Members/Purpose: Professional society of mathematicians and educators
that promotes the interests of mathematical scholarship and research.
Holds institutes, short courses, and symposia to further mathematical
research; compiles statistics. Maintains biographical archives, offers
placement services, compiles statistics.
Journals/Publications: Abstracts of papers presented to the AMS,
*Assistantships and Fellowships in the Mathematical Sciences, Bulletin of the
AMS, Combined Mathematical List, Current Mathematical Publications,*
Employment information in the mathematical sciences, *Journal of the
American Mathematical Society, Leningrad Mathematical Journal,
Mathematical Reviews*

Association of State Supervisors of Mathematics
c/o Wesley L. Bird, President
P.O. Box 480
Jefferson City, MO 65102-0480
cbmsweb.org/Members/assm.htm
Members/Purpose: Individuals serving on the education agency staff of
any U.S. state or possession, the District of Columbia, the U.S.
Department of Education, or any Canadian province or territory.
Promotes high standards in the teaching of mathematics; encourages
interest in mathematics and its teaching. Facilitates exchange of ideas
and information among members; promotes cooperation among
educational agencies. Identifies the needs of the future and makes
recommendations for improving mathematics education.
Journals/Publications: ASSM newsletter

Association for Women in Mathematics
11240 Waples Mill Rd., Suite 200
Fairfax, Virginia 22030
awm-math.org
Members/Purpose: Mathematicians employed by universities, government,
and private industry; students. Seeks to improve the status of women in
the mathematical profession and to make students aware of
opportunities for women in the field. Membership is open to all
individuals regardless of gender.

Journals/Publications: *Newsletter: Association for Women in Mathematics, Directory of Women in the Mathematical Sciences, Careers for Women in Mathematics, Careers That Count, Careers in Mathematics, Profiles of Women in Mathematics: The Emmy Noether Lecturers*

Conference Board of the Mathematical Sciences
1529 Eighteenth St. NW
Washington, DC 20036
cbmsweb.org

Members/Purpose: Is an umbrella organization consisting of fifteen professional societies, all of which have a primary objective to increase knowledge in one or more of the mathematical sciences. Its purpose is to promote understanding and cooperation among these national organizations so they work together and support each other in their efforts to promote research, improve education, and expand the uses of mathematics. The fourteen professional societies include AMATYC (American Mathematical Association of Two-Year Colleges), AMS (American Mathematical Society), ASA (American Statistical Association), ASL (Association for Symbolic Logic), AWM (Association for Women in Mathematics), ASSM (Association of State Supervisors of Mathematics), BBA (Benjamin Banneker Association), INFORMS (Institute for Operations Research and the Management Sciences), IMS (Institute of Mathematical Statistics), MAA (Mathematical Association of America), NAM (National Association of Mathematicians), NCSM (National Council of Supervisors of Mathematics), NCTM (National Council of Teachers of Mathematics), SIAM (Society for Industrial and Applied Mathematics), and SOA (Society of Actuaries).

Convention/Meeting: Semiannual council meeting

Mathematical Association of America
1529 Eighteenth St. NW
Washington, DC 20036-1358
maa.org

Members/Purpose: College mathematics teachers; individuals using mathematics as a tool in a business profession.

Journals/Publications: *American Mathematical Monthly, College Mathematics Journal, Mathematical Association of America Mathematics* magazine

Math/Science Network
Mills College
5000 MacArthur Blvd.
Oakland, CA 94613-1301
expandingyourhorizons.org
Members/Purpose: Mathematicians, scientists, counselors, parents, community leaders, and representatives from business and industry who are interested in increasing the participation of girls and women in mathematics, science, and technology.
Journals/Publications: *Broadcast/Beyond Equals: To Promote the Participation of Women in Mathematics, Expanding Your Horizons in Science and Math: A Handbook for Conference Planners*

National Council of Supervisors of Mathematics
P.O. Box 150368
Lakewood, CO 80215-0368
mathforum.org/ncsm
Members/Purpose: Supervisors of curriculum and personnel in mathematics departments at the elementary, secondary, and college levels. Seeks to develop solutions to problems in all areas of mathematics supervision and curriculum development; provides a forum for the exchange of ideas and current research results. Works with other groups to improve the teaching of mathematics.
Journals/Publications: Membership directory, newsletter

National Council of Teachers of Mathematics
1906 Association Dr.
Reston, VA 20191-1502
nctm.org
Members/Purpose: Teachers of mathematics in grades K–12, two-year colleges, and teacher education personnel on college campuses.
Journals/Publications: *Arithmetic Teacher, Journal for Research in Mathematics Education, Mathematics Teacher, National Council of Teachers of Mathematics—Yearbook*, NCTM news bulletin

National Education Association
1201 Sixteenth St. NW
Washington, DC 20036-3290
nea.org/index.html

Members/Purpose: Professional organization and union of elementary and secondary school teachers, college and university professors, administrators, principals, counselors, and others concerned with education.

Journals/Publications: *ESP Annual, ESP Progress,* handbook, *NEA Today, Almanac of Higher Education, NEA Higher Education Advocate, Thought and Action*

School Science and Mathematics Association
The Ohio State University
238 Arps Hall
Columbus, OH 43210-1172
ssma.org

Members/Purpose: Science and mathematics teachers at elementary through college levels and persons involved in teacher education. Objectives are to facilitate the dissemination of knowledge in mathematics and the sciences; to encourage critical thinking of knowledge in mathematics and the sciences; to encourage critical thinking and use of the scientific method; to emphasize the interdependence of mathematics and the sciences in education, research, writing, and curriculum development; to provide the means for dialogue among teachers of mathematics and the sciences; to identify and help solve problems common to science and mathematics education.

Journals/Publications: *School Science and Mathematics, School Science and Mathematics Association Convention Program,* newsletter

Society for Industrial and Applied Mathematics
3600 University City Science Center
Philadelphia, PA 19104-2688
siam.org

Members/Purpose: Mathematicians, engineers, computer scientists, physical scientists, bioscientists, educators, social scientists, and others utilizing mathematics for the solution of problems. Purposes are to promote research in applied mathematics and computational science; further the application of mathematics to new methods and techniques useful in industry and science; provide for the exchange of techniques useful in industry and science; provide for the exchange of information among the mathematical, industrial, and scientific communities.

Journals/Publications: *CBMS-NSF Regional Conference Series in Applied Mathematics, Classics in Applied Mathematics and Proceedings, Frontiers in*

Applied Mathematics, membership list, review, *SIAM Journal on Applied Mathematics*, *SIAM Journal on Computing*, *SIAM Journal on Control and Optimization*, *SIAM Journal on Discrete Mathematics*, *SIAM Journal on Mathematical Analysis*, *SIAM Journal on Matrix Analysis and Applications*, *SIAM Journal on Numerical Analysis*, *SIAM Journal on Scientific Computing*, *SIAM News*

Women and Mathematics Education

Pat Frey, WME Treasurer
24 Westminster Rd.
Buffalo, NY 14224
wme-usa.org

Members/Purpose: Individuals concerned with promoting the mathematical education of girls and women. Serves as a clearinghouse for ideas and resources in the area of women and mathematics. Establishes communications for networks focusing on doctoral students, elementary and secondary school teachers, and teacher educators. Encourages research in the area of women and mathematics, especially research that isolates factors contributing to the dropout rate of women in mathematics. Emphasizes the need for elementary and secondary school programs that encourage involvement in mathematics by females.

Journals/Publications: *Women and Mathematics*, *Women and Education*

7

Jobs Using Math as the Primary Skill

Mathematicians work in many different fields under many different job titles, but today, as in the last several decades, four job titles generally describe the specialists whose work involves the most concentrated daily mathematical activity. These are the **actuary, mathematician, statistician,** and **operations research analyst.**

If you want to focus on mathematics on a daily basis, the careers described in this chapter provide that degree of involvement. In the past, many career guides treated these jobs as the only positions for math majors who wanted to work primarily as mathematicians in their daily work. The variety of careers presented in this and later chapters demonstrates that there are many other choices in which you can apply your mathematics skills daily.

Just because a person is talented at math doesn't necessarily mean that he or she wants to live, breathe, and dream it. Some mathematicians do, while others may simply want to use their gifts to earn a living. Regardless of your individual preferences at this point, we want to give you as much information as possible so you can make a more informed career choice that will fit your specific aptitudes, skills, and needs. We will try to provide a balanced picture of these four traditional math occupations by describing both the pluses and minuses of each job, as well as information about the kinds of employers and job sites where these occupations can be found.

In every case, in each of the positions that follow, and in others suggested later in this chapter, your math degree will be the most important hiring qualification. Your math expertise would be utilized in these jobs every day. It would figure as a major influence on your career beginning with your first interview and would be basic to your advancement in any of these careers.

The job ad that follows is for an entry-level statistician and illustrates the emphasis that prospective employers place on your knowledge of math.

Statistician/Project Analyst: The qualified candidate will learn all phases of the model building process, including submitting SAS jobs, generating distributions, and generating reports for customer presentations (i.e., industry tables/performance charts). Requires a BA/BS degree in statistics, mathematics, economics, computer science, or related field. Experience in data manipulation and model building using statistical or related software is necessary. Knowledge of SAS and MS Office applications is essential. Must possess analytical skills to evaluate, understand, and interpret data from both our organization's and the customer's perspectives.

In considering an employment opportunity such as this, you need to have some concrete idea of your own value to the employment market. In 2003, entry-level salaries for new graduates with a bachelor's degree in mathematics averaged $40,512 a year, according to a survey by the National Association of Colleges and Employers. Offers to recent master's degree graduates averaged $42,348, and offers to recent Ph.D. graduates were averaging $55,485. These salary levels are as much as 10 percent to 20 percent higher than those for new graduates with business, communications, or education majors, and indicate the relative value of a mathematics education in the job market today.

If the expression "Money talks!" has any meaning for career counselors, it's a sure sign that the graduates of a particular major are in demand. There are two principal reasons for this demand: one is supply, the other is needed skills. Look around your math department and at some of the other departments on campus. You'll see one reason you're so much in demand—when the supply is thin, the price goes up. There are fewer math majors, because the work is not easy and there are also demanding examinations and professional requirements all along the way in many of the math-related professions.

The other reason is more complex and is of importance to your entry into your first job and your career progress. Let's consider the quoted starting salary for your major and ask, "Which of your fellow graduates from the variety of majors at your school is ready to sit down at a desk or conference table, or go out on a consulting call and immediately accomplish something that is of basic importance to the employer?" One of the most qualified is the math major. Now it becomes clear why the starting salaries for a math major are so high.

The point we are stressing here is that, of the many college majors offered, math is one of the few that actually delivers graduates who have specific skills, discipline, and knowledge that can be applied immediately, with minimal employer-supplied training. This job readiness makes it easier for undergraduate math majors to answer the question asked so frequently of them: "What are you going to do with that?"

Let's take a look now at each of the four traditional career options for math majors to see what they have to offer, specifically for you.

Actuary

Actuaries are experts in mathematical calculation and analysis, and assessment of probabilities and financial risk. For this reason, about 60 percent of all actuaries work in the insurance industry, with the rest working in financial credit and investment, pension funds, econometric forecasting, and various areas of management, including financial management, public relations, and consulting. Some are also employed by securities and commodities brokers, and by government agencies.

If you haven't appreciated the role of the actuary in today's business community, you soon will. Making your own insurance payments when you graduate from college may expose you to issues in which actuaries have played a role. Why do young women pay less for the same coverage on their cars than young men do? How can term life insurance be less expensive than whole life insurance policies and still be a good thing? How can companies sell such enormous life insurance policies so cheaply at airports? The answers to these and many other questions about financial cost, investment, risk, and probabilities are in the work done by actuaries. They make a crucial contribution to any insurance company that is attempting to stay competitive with other similar companies and yet maintain insurance rates that will supply the necessary cash reserves to pay claims made against the organization.

Most actuaries specialize in a particular area—life insurance, health insurance, or property or liability insurance, for example. There is a very large and growing field of pension actuaries who manage the risk involved in retirement planning.

Actuarial jobs involve lots of math. Math skill is critical, but actuaries are also problem solvers who need to put their talents to work in a number of different settings. They need to be good listeners in order to understand the issues involved and be well read to keep up on current trends and issues in business, the social sciences, law, and economics. They have a professional interest in news stories such as the risks of secondhand smoke, the length of hospital stays for pregnancy, studies of teenage depression, or any of the myriad social and cultural factors that can affect health and longevity and risk management in insurance.

The following is a description of an entry-level actuarial job. The employer is a major life insurance annuity business that stresses the factors we have described.

Actuarial Student/Assistant Actuary: Responsibilities include preparing quarterly reserve information, maintaining segmented asset models, providing various projections related to the company's assets, preparing mortality studies, calculating reserve factors, and working on special projects as needed. Requirements: BS in mathematics, strong computer skills, high level of general information, impeccable detail orientation, and strong verbal and written skills. Competitive salary and comprehensive benefits.

Mathematical expertise may be crucial for this job, but you can readily see that actuaries also need to be well rounded. While they may receive comparatively high salaries, they also earn them. To be successful, they must have a combination of math knowledge and general knowledge in regard to business and social trends.

Like other members of a management team, actuaries must be good communicators. The truly valuable actuary is the man or woman who can take the complicated situations and variables to project probabilities of sickness, death, injury, disability, unemployment, retirement, or property loss and explain these in terms that nonactuaries on the management team can understand.

Few professional groups are more attuned to the critical issues of our changing culture than actuaries. Let's take AIDS, for example. Think of the increasing numbers of people affected by this disease and the almost daily changing projections for their health and longevity as new treatments and medications come on the market. Actuaries dealing with these kinds of health issues are among the best informed professionals in the world regarding state-of-the-art AIDS treatments and their degrees of success and failure.

This leads to another important point. Actuaries create and maintain valuable statistics that help organizations project the results of current trends for the future and offer the best services to their customers at the most competitive prices. Many students we talk with protest that they don't want a "desk job." In fact, not many jobs for college graduates could be classified strictly as desk jobs. Actuarial jobs often involve significant travel and time spent working away from the office.

An actuarial career requires multiple skills and abilities and is part of the service, or "helping," professions. Without actuaries, insurance companies would have no idea what to charge policyholders for the coverage they want and—just as important—might not be prepared to pay out claims for the damages that the insurance was purchased to indemnify.

Following is a short checklist of skills and qualities that you need to be a successful actuary. If most of these are true of you, you might seriously investigate the actuarial field.

- Do you have a strong interest in business as well as mathematics?
- Do you have a strong background in business, or are you willing to acquire more education and training to get it?
- Do you truly enjoy problem solving, looking at problems from different perspectives, and finding new and creative ways of solving them?
- Do you read the newspaper, watch the news, and generally stay abreast of what goes on in the world? Is your level of general information high?
- Do you have superior math skills? This question applies to every job in this chapter, but it's particularly applicable for actuaries. Math is the very heart and soul of this job, and you need to ask yourself not only "Am I good?" but "Is math something I can see myself doing for a large part of every day?"

Actuarial applicants may have degrees in mathematics; actuarial science; statistics; or business, finance, or accounting. Computer skills are increasingly important, and actuaries need programming expertise and knowledge of spreadsheets and databases, plus standard statistical analysis software (SAS).

Requirements for Professional Status

The Society of Actuaries and the Casualty Actuarial Society give a series of four exams, sponsored jointly, which test competence in probability, calculus, statistics, and other areas of mathematics. Following this group of four exams, each society gives additional exams for higher levels of professional designations. It is possible to begin to take these exams while still in college. The preparation is time consuming and rigorous, but having already passed at least one of the actuarial exams is a strong advantage in applying for a job.

Once an actuary is on the job, he or she will continue to take the professional examinations, usually with the support of the employer, who may provide on-the-job training and some of the time needed for study, and may also pay for the examination fee and the study materials. The time needed for preparation for an actuarial exam is considerable and comparable to other professional exams, such as those taken by public accountants and lawyers. Some employers give awards or merit pay increases with each exam level that is successfully completed.

Actuaries are expected to complete the series of exams at fairly regular intervals as they progress in their careers. Usually an actuary will be able to complete all the exams for the Associate level in four or five years and the Fellowship level, within a few years more. Pension actuaries are required to have additional experience and pass specific examinations in order to be enrolled in the federal government's Joint Board for the Enrollment of Actuaries.

Mathematician

Mathematicians work in applied mathematics or in theoretical—sometimes referred to as "pure"—mathematics. An aura of respect and fascination has been associated with practitioners of mathematics since ancient times. It is one of our oldest sciences, and throughout its history has continued to be challenging and deeply engaging to many.

Evidence of popular interest in mathematics can frequently be seen in modern media. In recent years, two bestselling books on the same mathematical mystery have been published—*Fermat's Enigma* and *Fermat's Last Theorem*. Pierre de Fermat was born in Toulouse, France, in 1601. In 1637, scribbling in the margin of a book, he indicated that he had solved a celebrated theory of numbers but lamented that the book's margin did not contain enough space to record the solution! For more than 300 years, mathematicians labored to find this solution.

Finally, in 1993, Andrew Wiles of Princeton announced that he had solved the problem, only to have someone discover that his solution was flawed. Although he had already worked on the problem for seven years, Professor Wiles went back to work for another year, to firmly establish his solution, thereby also reestablishing his reputation among mathematicians and simultaneously charming the media and many in the general public.

You may be familiar with the movie, *Contact*, starring Jodie Foster, who plays the character of an astronomer who is devoted to listening for signs of extraterrestrial life in the universe. When she eventually makes contact with another planet, it is through prime numbers. Her character is not the least bit surprised by this, for as she says, "Mathematics is the universal language!" While this movie is fiction, the idea of the use of numbers and mathematics as a possible means to communicate with an extraterrestrial life form has been present since before the earliest days of space exploration.

A third indication of popular interest and the multisided place of mathematics in our lives is represented by another book, *The Number Sense: How the Mind Creates Mathematics*, by Stanislas Dehaene. This is an important study that says humans can count before they can speak. Dehaene, who has received numerous awards in Europe and America for his work, sees mathematics not as something we acquire from someplace else but as something inherent in our mind. He goes on to illustrate how culture, and especially language, can enhance or impede that mathematical sense.

Obviously, the work of mathematics is perennially ongoing, and there is so much more to learn. Mathematicians do research in fundamental or theoretical mathematics, or in the application of mathematics to areas such as economics, social science, engineering, and other fields. They try to learn more about what math has to tell us or apply direct mathematical solutions to various fields. All the areas of math are involved, including algebra, geometry, number theories, logic, and topology. Generally, the work of mathematics is either applied or theoretical, but the work done in many math occupations blurs the distinction between the two.

Essentially, those working in theoretical math occupations are seeking to enhance our understanding of mathematics. Their concern is not how increased understanding might immediately affect our practical lives but simply to know more. Theoretical mathematicians have given us many great achievements that have had profound effects in science and engineering.

Those working in the field of applied mathematics are using their math skills to develop mathematical models or computational methods that can be used to solve real-world problems—often in business, government, science, engineering, and a host of other disciplines. Sophisticated advertising agencies are increasingly developing mathematical models to account for the variables that affect a product's life cycle and the effects of increased expenditures for advertising or promotion.

The following job announcement is for a recent math graduate with some professional or internship experience in using data management:

Here's an example of how applied mathematics made headline news. In November 1996, *The Boston Globe* banner headlined "$200 Million Error—in Your Favor." Burt Feinberg, a mathematician for the Massachusetts State

> **Mathematician/SAS Analyst:** Provide technical support to an epidemiological research group performing large, data-intensive studies of the health of military personnel. Requires at least one year of experience in data management and analysis using SAS. Must be able to perform frequency and univariate analyses on study data. Working knowledge of standard PC software. Formal training in statistics, biostatistics, or epidemiology a plus.

Rating Bureau, a division of the state's Division of Insurance, discovered that consumers had been overcharged since 1991, because of the manner in which industry expenses had been calculated in annual rate cases. As quoted in the newspapers over the many days this story ran, "This is an extremely complicated, arcane thing. From a consumer's standpoint, it's not an understatement to say that Burt's a hero on this issue."

What's algebra worth? In just that one instance, it was worth $200 million to the people of Massachusetts. Mr. Feinberg was profiled as being born to working-class parents in Long Beach, New York, in an environment where playing with numbers was a way of life. "I guess I was just always ahead in math. Mom would feed me simple arithmetic problems. I saw it as fun." His boss reported in the newspapers that, in addition to being good at math, Mr. Feinberg was also good at "stepping back and looking at the big picture."

Mathematicians use computers extensively in model making, analyzing and correlating relationships, and processing large amounts of data. The most valuable hires in both theoretical and applied math settings are those individuals who have learned a number of relevant software packages and have been involved in creating programs.

Later in this chapter, you'll see that mathematicians have a variety of job titles. The reason for this is that both theoretical and applied math occur in many settings. In these varied settings, mathematicians may be called researchers, analysts, specialists, and other titles that are more reflective of the field they are working in. A cryptologist (the branch of mathematics that deals with designing codes to transmit secret information), an agricultural economist, and an operations researcher are all mathematicians.

Statistician

The Dictionary of Occupational Titles, or *DOT*, is written by economists and other specialists employed by the U.S. Department of Labor's Bureau of

Labor Statistics and provides remarkably accurate and useful information. Here's *DOT*'s description of a statistician:

> **Statistician, Mathematical.** Conducts research into mathematical theories and proofs that form basis of science of statistics and develops statistical methodology. Examines theories, such as those of probability and inference, to discover mathematical bases for new or improved methods of obtaining and evaluating numerical data. Develops and tests experimental designs, sampling techniques, and analytical methods, and prepares recommendations concerning their utilization in statistical surveys, experiments, and tests. Investigates, evaluates, and prepares reports on applicability, efficiency, and accuracy of statistical methods used by physical and social scientists.

This description doesn't begin to suggest the variety, interest, and scope of the statistician's possible work responsibilities or settings. In the government and private sector, statisticians analyze consumer prices, employment patterns, and population trends. The results of their work will have a major impact on public policy, the administration of many social programs, and the practices and policies of many private businesses.

But that's not all. Scientific research of all kinds uses statistical information and employs the mathematicians who are qualified to handle that material. The task might be radiocarbon dating of volcanoes or biochemical work in new drug trials. It could involve new agricultural techniques or human behavior studies that help us understand what we do and why we do it.

Statisticians may be employed in private industry to ensure that the company is neither under- nor overstocked with inventory. Statisticians may be asked to predict consumer behavior, maintain quality of production, or ensure the viability of retirement investment accounts. The list goes on and on, and the applications of statistical mathematics are increasing every day in our competitive economy.

In the work of statisticians, you can see once again that mathematics and those who practice it are on the cutting edge of the issues and concerns of our society. The search for new and improved medical therapies for cancer, Alzheimer's disease, and cystic fibrosis, and the work of many of the medical challenges of the twenty-first century are being performed by teams of scientists that include statisticians. Likewise, in our increasing concern for the environment, we need the skill of statisticians to help us understand and see the impact of human behavior on the life of this fragile planet.

Such concerns occupy the working time of many statisticians. But statisticians are also working in many other fields with smaller but equally important questions, such as those of marketers (What are people buying at the grocery store?), advertisers (What is America watching on TV this week?), developers (Where should we site this new fast-food franchise?), and social scientists (How many emergency room admissions are the result of gang violence?).

Ask yourself the following questions to test your interest in being a statistician:

- Are you good at and do you enjoy both math and computers?
- Do you like working with real-life problems?
- Are you successful and do you enjoy working as part of a team? Much of the work of a statistician is done as part of a larger group effort.
- Are you a good listener? Statisticians frequently need to talk with people who don't understand statistics but need statistical information. To understand people's needs, you need to be a good listener.
- How do you feel about a steady diet of new problems, new challenges, and learning new things?

If you answered these questions in the affirmative, it's a good indication that you could be a candidate for the jobs offered to statisticians.

Operations Research Analyst

Our fourth category of jobs that use mathematics as a primary skill is operations research analyst, sometimes referred to as management science analyst. This is another fascinating job category where your mathematics profession can involve you with knowledge of almost any field of human endeavor.

For a dramatic example of what some operations research analysts do, let's take a look at any major metropolitan airport such as Chicago's O'Hare International. The video monitors showing arrivals and departures of aircraft from all over the world attest to a staggering amount of scheduling. Add to that the incredibly complex system of baggage handling and routing as passengers for different flights check in at the same counter within minutes of each other. Think, too, of food and beverage preparation and stocking requirements for all the aircraft. Most important are the frequent on-ground maintenance checks and quick repairs that must be done to keep the aircraft flight worthy. Consider the logistics involved in making sure that all those parts are available, your favorite food and beverage are on board, and your luggage arrives at your destination when you do. Running any complex organization requires massive amounts of coordination of people, machines, and materials.

Operations research analysts help large organizations mount and run systems such as in our airport example. They do this by applying mathematical concepts to organizational problems, generating a number of possible solutions, and then choosing those that meet the organization's goals, finances, mission, and philosophy.

The following ad for an entry-level operations research analyst will serve as an excellent example of what they do, because the industry (small-package transportation market) advertising the job shares many similarities to our airport baggage example:

A rapidly growing leader in the small-package transportation market is seeking candidates in business planning and operations research. Duties include developing and utilizing mathematical models and analytical tools to improve current operation and reduce operating costs, create and implement new tools to monitor and improve performance, provide operational support for planning and maintaining/implementing existing and new services, enhance/implement operational research techniques.

This position requires a BS degree in mathematics or operations research. Must have excellent written and spoken communication skills, 1–2 years experience including security operations, good working knowledge of professional software, including SAS, plus Microsoft Office, and be able to work independently at problem solving in a dynamic work environment.

Predicting how systems will work involves mathematical modeling, extensive use of the computer, and consultation with other people, especially people on-site who will be involved in making these systems operational. Computer use, listening skills, and teamwork. Sound familiar? Like every other job type mentioned in this chapter, operations research analysts share these common needs.

Problem solving begins with a needs assessment, which begins by talking to working professionals and trying to understand the problems and nature of their work. Perhaps the project is to assess economic order quantities for materials purchases. Overordering means risking damage, loss, or stock becoming out of date; underordering means risking a stockout and/or a production shutdown.

Next, an operations research analyst would analyze the problem and isolate all its elements, ensuring that he or she understands and has complete information about each one. For example, a comprehensive list of acceptable alternate shipping methods for transporting parts and a detailed list of direct computer links to the parts suppliers might be needed.

The analyst will then begin to design a series of analytical techniques using a number of mathematical models to construct a system that solves the problem at hand. These methods might include simulations, linear optimization, networks, waiting lines, and game theory.

Computerized analyses generally involve significant work with databases. Operations research analysts need to master database collection, management, and programming to be effective. Many of the mathematical models employed by operations research analysts are large and multilayered, and use computers to process.

Working Conditions

Mathematicians in the traditional math jobs generally work regular hours in an office environment, although there may be frequent team meetings and project committee work. None of these jobs are deskbound; some, including actuaries, may travel to branch offices or offices of clients. Frequently, there are deadlines to be met, and that means working under pressure. These jobs are for management-level, salaried employees, and it is not uncommon to work more than a forty-hour week to accomplish specific tasks or during certain times of the year. Because new information is vital to these jobs, there is typically frequent opportunity for professional development, which may include traveling to seminars and conferences.

Training and Qualifications

Strong math and computer backgrounds are essential for all of the jobs listed in this chapter. These positions require you to think logically, work well with people, and have good oral and written communication skills. Each job category, however, presents its own unique set of training and qualification demands.

Actuary

Superior math skills are required for the actuary, as for all the jobs in this chapter. It would not be uncommon during the interview process for the employer to discuss mathematical issues with you but, perhaps, not on the first interview. If you return for a second interview, you may be shown work in progress and asked to comment on it. You might be asked to spend some time on a sample problem or case study involving mathematical processes. So your math must be top-notch.

Your transcript and performance in your major courses will be of real interest to prospective employers, since many of them have taken the same courses. All of these interview procedures help the employer understand your strengths and weaknesses and have a richer appreciation for how you might be utilized in the organization.

The rewards of a career in actuarial work (and there are many rewards) are directly related to the passing of the many levels of qualifying examinations to achieve professional standing. All professional actuaries begin by taking the same four initial examinations and then, following specialization, go on to take their own specific series of additional examinations in their professional specialty. It's not unusual for some college students to have taken and passed one or two of these exams prior to graduation. Generally speaking, it takes from five to ten years to pass the entire series. With the passage of different exams, you move from actuary trainee, to actuary, to actuary associate.

Mathematician

Many jobs for mathematicians require advanced degrees, but there are some jobs available for applicants with only a bachelor's degree. In the federal government (the largest employer of math majors), you must have a four-year degree with a math major for entry-level positions as a mathematician. Recommended undergraduate courses include calculus, differential equations, and linear and abstract algebra. Beyond this, courses in mathematical analysis, topology, numerical analysis, probability theory, and statistics are also worthwhile.

Taking mathematics courses in other disciplines at your college would be very advantageous if you can meet the prerequisites and have an interest in the subject matter. Look into the math offerings in the computer, engineering, economics, and physical and social sciences departments.

In the private sector, too, positions will be available for every degree level; however, the entry-level positions may be titled something other than mathematician. You may be called a researcher, programmer, systems analyst, or systems engineer. Generally, the full title of mathematician is reserved for those holding advanced degrees.

Statistician

Although many employers are looking for advanced degrees, math majors can find employment with their bachelor's degree as statisticians. Many of you will have taken a significant number of the courses offered in your math department at college. Those will be important, of course, but your employer will also be interested in what and how many statistics courses you took as well. Try to take as many specific courses in statistics as your math department offers.

Don't neglect the opportunity to take additional hours of quantitative courses in any other discipline your college might offer. For example, there may be statistics courses in engineering, education, pre-med, social science, or biology. Depending on your interests and the prerequisites for those courses, you might want to try your hand at some of those. Statisticians work in a variety of settings, and such course work helps extend your qualifications.

For federal positions, job candidates will need to document at least six semester hours in statistics courses and nine semester hours in other math courses. Higher-level positions with the government have commensurately greater demands for advanced math courses.

As for the positions described in this chapter, employers seeking statisticians want good communicators, and those hiring in the private sector want job candidates who also have a solid working knowledge of business, economics, and management practices.

Operations Research Analyst

As an entry-level employee, you will begin your career doing routine assignments under the supervision of an experienced supervisor who most likely will have an advanced degree. Many employers prefer operational research analysts to have at least a master's degree. There are entry-level positions in this field, but you'll have to work hard to prove yourself, as the premium candidates generally have advanced degrees. Your entry-level title may be coordinator or project manager.

As employees with a bachelor's degree advance with their employer in operational management, they will usually attend university classes to gain their master's degree, often at the employer's expense or at least with partial tuition support. To move ahead in your career, advanced study is more a requirement than an option. The great thing about working on a master's degree while you are employed is that, although it certainly impacts your social life and sleep, your job gives you a useful day-to-day laboratory of experiences that can make the learning come alive.

Your résumé and interviews should demonstrate your continuing professional development, and ability to think logically and work well with people. You should provide some examples of your written work, as your communication skills during your interview will also be judged. Your résumé should also include a section devoted to your computer skills, in an organized grouping of the hardware, systems, and software with which you are proficient.

The following ad is for a project analyst, a common entry-level job title in operations research, that touches on many common employer demands:

> **Project Analyst:** The Dynamic Group is a $6.2 billion firm owned by the XYZ Corporation, parent company of ABC Airlines. We are the leading provider of business decision–support tools for the transportation and travel industry. Using innovative operations research techniques and cutting-edge technology, we provide creative solutions for our clients' toughest problems. Success in the role of project analyst requires a BS or better in mathematics or operations research, industrial engineering, computer science, or equally analytical discipline. Analysts are the primary point of contact for our clients and perform a range of functions. They are responsible for implemention of our software at clients' sites, preparing and making presentations to all levels of clients' corporate employees and management, training users, preparing client deliverables including technical writing, carrying out troubleshooting, and providing ongoing customer support. This position requires strong problem-solving and communication skills, a teamwork ethic, and the ability to balance multiple tasks, deadlines, and clients. Approximately 30% travel is required.

Earnings

Salary is a funny aspect of a job. If you hate your work tasks, a high salary may still be reward enough to keep you in a job that you don't enjoy, because you have the resources to utilize your nonwork time as you wish; you can compensate. If you are unhappy with the job *and* your coworkers, however, it has been our experience that money is unimportant and people will do anything to change jobs, including taking a pay cut.

Actuary

Salaries for entry-level actuarial jobs averaged about $46,991, according to a Life Office Management Association survey in 2003. Associate actuaries received an average of $99,446, while actuaries at the highest levels without managerial responsibilities received $104,235, on average. Those with managerial responsibilities received much more, with salaries and other benefits varying widely.

Mathematician

The National Association of Colleges and Employers reported in a 2003 survey that the average starting salary for graduates with a bachelor's degree in mathematics was $40,512 per year and $42,348 for those with a master's degree. New job candidates with a Ph.D. averaged $55,485.

The median salary for mathematicians in the United States in 2002 was $76,470. The lowest 10 percent earned less than $38,930, and the highest 10 percent earned more than $112,780. Salaries with the U.S. government were slightly higher.

These are high starting salaries. Stop by your career office, and look at their estimates of starting salaries for business majors, accountants, marketing and advertising majors, and others, and you'll find the average starting salary in math is among the highest.

Statistician

According to the U.S. Department of Labor's Bureau of Labor Statistics, the median annual earnings of statisticians in the private sector in 2002 were $57,080, with the lowest 10 percent earning $30,308 or less, and the highest 10 percent earning more than $91,680. Federal government statisticians averaged $75,979 in 2003, and this is an average of all statisticians' salaries at nonsupervisory, supervisory, and management levels.

Although these are median and average salaries, they give you a sense of the pay levels. Statisticians who earned advanced degrees received even higher salaries.

Operations Research Analyst

Because of the diversity of the work of operations research analysts, they are employed in many industries, and salaries vary considerably. However, median and average salary levels are available. Median annual earnings are quoted from the *Occupational Outlook Handbook* as being $56,920 in 2002, in general industry. Of these, the highest 10 percent earned more than $92,430, and the lowest 10 percent earned less than $34,140. Operations research analysts working in all levels of jobs for the federal government made more and averaged, in nonsupervisory, supervisory, and managerial positions, $83,740 in 2003.

Employment Outlook: Actuary

Jobs for actuaries are expected to grow at about the average rate through 2012. If there are significant changes in the insurance industry in the coming decade, there may also be significant changes in this job market. New forms of insurance, and especially annuities, would almost certainly create at least some new jobs. The health-care industry also faces changes, and these may affect the actuarial job market as well. Job opportunities are concentrated in metropolitan areas, and about a third of all actuarial jobs are in Connecticut, Illinois, New Jersey, and New York.

Employment Outlook: Mathematician

Employment for mathematicians is not expected to grow as fast as the average in the next decade. The already keen competition is expected to increase, and the number of jobs for mathematicians is expected to shrink somewhat, according to the U.S. Department of Labor's Bureau of Labor Statistics.

Earning a bachelor's degree in mathematics does not generally qualify you for most jobs as a mathematician. However, if you can expand your background to include computer science, electrical or mechanical engineering, or operations research, you'll find this opens up opportunities in industry. Bachelor's degree candidates meeting state teacher certification may become eligible for high school mathematics teaching positions.

Master's degree candidates in mathematics face strong competition for jobs in theoretical research, but jobs are more numerous in applied mathematics and related areas, such as operations research, engineering, and computer science.

Expectations by employers for Ph.Ds. are expected to increase, except for jobs with the U.S. government. The *Occupational Outlook Handbook* advises that all job applicants with advanced mathematics degrees and who also have additional degrees in engineering, computer science, or other related areas will likely have better job opportunities in related occupations.

At this time, about 20,000 mathematicians work in universities and colleges, and another 2,000 are employed in industry. The federal government employs a large number of mathematicians in the Department of Defense and a somewhat smaller number in the National Aeronautics and Space Administration. In addition, untold numbers are working in related areas where their mathematic skills are a lesser part of their work responsibilities and are not reflected in their job titles. Because the number of doctorates earned in mathematics continues to exceed the number of jobs available in theoretical research, many of these graduates will have to look to industry and government for jobs.

Employment Outlook: Statistician

The outlook here is an interesting combination of supply and demand. While actual numbers of jobs may not grow faster than the average for statisticians, a favorable job outlook is still predicted for those with the degree and talent.

Both the government and private industry will remain strong employers, with private industry placing special emphasis on those with advanced degrees. Car manufacturers, drug companies, engineering firms, and producers of chemicals and foodstuffs all present possibilities, as does the increasing use of statisticians by marketing firms and advertising agencies to forecast and track sales, and solve product positioning problems.

Employment Outlook: Operations Research Analyst

Your best bet in this field is to get a job as a research assistant of some sort (job titles will vary from employer to employer), where the decided emphasis in your job is on quantitative analysis. Learn as much as you can, and keep an eye out for an opportunity to advance your education. The number of operations research or management science graduates does not meet the number of positions available. Even so, employers will place a special premium on those with master's and doctoral degrees.

Strategies for Finding Jobs

Strategies to use for finding jobs that use math as a primary skill, such as actuaries, mathematicians, statisticians, and operations research analysts, depend on the supply and demand in these fields, the type of industry, and the geographic location. In these occupations, your math education is essential for breaking into the field and for continued growth.

Some common strategies exist for all four of these occupational areas. First, find out where the jobs are. If there is little demand within your preferred geographic location, you may need to consider relocating. Second, find the position that matches your mathematical skills and experience. Third, design a résumé tailored to the position description in the posting. And fourth, write a cover letter that focuses on the benefits of hiring you and describes the credentials that match the needs in the position you are seeking. If hiring officials are able to match the qualifications stated in your résumé and cover letter with their needs, then there is a good chance you will be asked to come in for an interview. During the interview, if the match between your qualifications and the employer's needs is further supported, then the final goal will be achieved—getting hired!

The best strategy is a proactive strategy, because only 20 percent of all jobs are advertised, which leaves 80 percent that are unadvertised. With a proactive strategy, you will use your research skills to find the unadvertised jobs. These unadvertised postings are referred to as the hidden job market. The key is finding these hidden opportunities in organizations that currently employ individuals who have a degree in mathematics.

Begin your search by thinking about networking from known contacts to new contacts. Start a personal and professional network of contacts. The known contacts include those students who are in your mathematics classes at college, the professors who teach mathematics courses, and the clubs and organizations on campus. Be proactive by searching the directories of employers by occupation, field, and geographic areas. Use the Internet to search for websites that offer addresses for job opportunities. Search the yellow pages of your local telephone directory. Pick up the phone and call organizations to find out what they have for opportunities and where they advertise when they do have openings. Use the skill that most mathematicians have— "research" all possibilities!

Check out the careers library resources, campus library, career office postings, mathematics department bulletin board, and professional associations. All of these physical locations may list websites as well. For instance, *Peterson's Job Opportunities for Business Majors* and *Peterson's Job Opportunities for Engineering and Computer Science Majors* include overviews of organizations, key statistics about the organizations, expertise/education sought by the organizations, international assignments, alternatives to assignments (internships), contact names, and organizations' websites for many of their listings in the areas of actuarial science, mathematics, operations research, and statistics.

At the end of this chapter, several associations and organizations are listed that not only have websites but also publish newsletters, journals, and online listings of positions, from bachelor's degree level to Ph.D. In addition, they offer hot tips on job search strategies. Review the online newspapers and journals for possible career positions. Find out if there are local, regional, or national job fairs being offered. Sometimes employers send their part-time and full-time listings directly to the math department and organizations on campuses. If you happen to find an internship listing that suggests the type of organization and skills you are seeking, then check with that employer to find out if they have full-time listings.

Following are some strategies and possible employers that would relate specifically to actuaries, mathematicians, statisticians, and operations research analysts.

Actuary

About 60 percent of all actuaries are employed in the insurance industry, while others work for firms providing services such as management or public relations, or in actuarial consulting firms. A small number of actuaries work for security and commodity brokers, government agencies, and computer software developers.

Actuaries most often specialize in life, health, or property and casualty insurance, while other actuaries specialize in pension plans. To learn various actuarial operations and phases, beginning actuaries often rotate between jobs, and they may move from one company to another in their early years to find growth and more responsibility. As reported in the *Occupational Outlook Handbook*, employment of actuaries is expected to grow more slowly than the average for all occupations; however, employment of consulting actuaries is expected to grow faster than for actuaries employed by insurance carriers.

Possible Employers

To find possible employers who have actuarial positions, use your research skills, and take a proactive approach to search for openings with insurance companies, educational institutions, government agencies, financial institutions, computer software companies, consulting firms, and public relations firms. Use telephone directories, the Internet, campus career services, and actuarial associations.

Mathematician

Because of the two broad classes in which mathematicians fall (theoretical and applied), a different job strategy must be used for each. The job search for mathematicians starts with a broad search, but it needs to be narrowed by developing a strategy. That strategy will be for you to decide:

1. Geographically, where do you want to work?
2. Choose the industry in which you would like to work:
   - Government (the federal government employs 75 percent of all mathematicians)—take a look on the Internet at the Office of Personnel Management for positions with the federal government at usajobs.opm.gov
   - Manufacturing (pharmaceuticals are the major employers)
   - Private sector (management and publication services, educational services, research testing, security and commodity exchanges, and employers such as insurance companies and banks)
3. Search the industries using various avenues, with a reactive approach (advertisements that are printed in newspapers, magazines, or journals) or proactive approach (search for employers that are in your geographic area but have not advertised that they have openings, by looking on the Internet, placing telephone calls, and contacting professional organizations that have their own job listings).

Possible Employers

Use some of the same job search strategies for mathematicians as for actuaries. Possible employers are found in industries such as government, manufacturing, education, and the private sector. Research your college's careers library resources, campus library, career office postings, mathematics department bulletin board on campus, and professional associations for mathematician job postings. Mathematics is the skill used and not the descriptive occupational title used in job listings as it was for statisticians and actuaries.

If you are using the Internet for your job search, you will find that, by entering the words *theoretical mathematician*, job titles such as analyst, technician, engineer, research associate, systems designer, and scientist will pop up on the lists. Entering the job title of *applied mathematician* brings up listings such as programmer, developer, finance associate, maintenance technician, quality assurance engineer, applied systems sales, architect, consultant, and business solutions specialist.

Statistician

Statisticians need a variety of skills, because it is not enough just to be good at statistics to succeed in this field. Communication skills, a good understanding of business and the economy, and the ability to explain technical processes to those who are not statisticians are all important to make you more attractive to any industry. After you are employed, you will find that advancement comes with experience gained in the field and a more advanced degree.

As reported in the *Occupational Outlook Handbook*, employment for statisticians is expected to grow little through the year 2012. Those with bachelor's degrees will discover that they will be finding jobs that don't have the title of statistician, especially in the areas of engineering, economics, biology, or psychology. To get a great start as a statistician, students will need to have a strong background in mathematics, engineering, or computer science. With this experience, graduates should have the best prospects of finding jobs, especially in the federal government, where approximately 25 percent of the jobs are found. Even then, competition will be strong. If you get state certification, there are opportunities to teach high school statistics. Look for full-time opportunities in private industry: pharmaceuticals, manufacturers, research and development businesses, and consulting firms.

As a statistician, you have learned the skills that can be applied to a good job search strategy, such as the following:

1. Decide where and how to gather the data (whether your job search is proactive or reactive).
2. Determine the type and size of a sample group (the industry).
3. Develop a reporting form (your contact list).
4. Implement your plan: network; call; write; e-mail; and search the Internet, directories, and your career services library.

Possible Employers

Where do statisticians work? What employers do you explore? Use your research skills, and take a proactive approach to search for openings in the field of medicine, psychology, engineering, business, manufacturing, and biology. Many positions in statistics in private industry and colleges and universities require a master's or a doctorate. The word *statistician* is often incorporated into the title, or it may be hidden altogether.

Operations Research Analyst

When employers are seeking to fill an operations research analyst position, they are seeking two primary skills: (1) the ability to think logically and work well with people, and (2) the ability to use a computer, which is the most important tool for quantitative analysis, and a background in programming. Employers also prefer applicants with at least a master's degree in operations research, industrial engineering, or management science, coupled with a bachelor's degree in computer science, economics, or statistics.

Where are the jobs? Because operations research analysts help organizations coordinate and operate in the most efficient manner by applying mathematical principles to organizational problems, the jobs are located in various industries, especially those that deal with the problems occurring in large businesses and government organizations. As reported in the *Occupational Outlook Handbook*, individuals with a master's or Ph.D. in management science or operations research should find good job prospects through the year 2012, despite projected slower-than-average employment growth. Some operations research analysts are generalists, and some specialize in one type of application. Sometimes their work is embedded in the work of economists, systems analysts, mathematicians, and industrial engineers. However, operations research analysts are employed in most industries.

An analyst begins working under the supervision of a more experienced analyst and is then assigned more responsibilities to design models and solve problems. He or she advances by becoming a technical specialist and working as a supervisor. With only a bachelor's degree, you can find opportunities as a research assistant in a variety of related fields that allow you to use your mathematical skills.

Possible Employers

Operations research analysts are employed by the federal government, engineering and management services firms, insurance carriers, financial institutions, computer and data processing services, telecommunications companies, and air carriers. As reported in the *Occupational Outlook Handbook*, 20 percent work for management, research, public relations, and testing agencies that do operations research consulting.

Possible Job Titles

Specific job titles frequently do not contain the words *actuary, mathematician, statistician,* or *operations research analyst*. The titles and jobs themselves expand or overlap into technician, consultant, engineer, associate, designer, and more, as you will see from the following list.

Actuary

The title of actuary is an earned professional designation similar to CPA, lawyer, or realtor, but you might encounter ads for specific jobs that list titles such as the following:

Finance actuary
Health actuary
International actuary
Investment actuary
Life insurance actuary
Long-term care actuary
Retirement/pension actuary
Risk-management actuary

Mathematician
Possible job titles include the following:

Accounting analyst
Applied mathematician
Applied systems sales
Contracts specialist
Director of information technology
Finance associate
Marketing associate
Operations research analyst—operations specialist
Product quality engineer
Program manager
Project analyst
Project/program analyst
Quality assurance engineer
Quality engineer
Research associate
Software engineer/programmer
Systems analyst/programmer
Systems level designer
Test technician
Theoretical mathematician

Statistician
Possible job titles include the following:

Assistant professor—statistics
Biostatistician
Business analyst
Clinical trials statistician
Department chair, public health sciences

Director of management science
Economics analyst
Epidemiologist
Financial analyst
Global manufacturing analyst
Manager of marketing science
Programmer analyst
Quality assurance analyst
Research associate
Risk management analyst
Senior analyst
Senior marketing analyst
Senior mathematical statistician
Senior researcher for survey methodology
Statistical analyst
Statistical programmer

Operations Research Analyst

Possible job titles include the following:

Principal analyst
Product marketing manager
Project manager
Quality and customer satisfaction consultant
Research analyst
Risk management manager
Senior financial analyst
Senior project analyst for strategic financial management
Systems analyst

Related Occupations

For each of the areas studied in this chapter, there are also a number of related jobs and job titles with which you will want to be familiar. Some of these follow:

Accountant
Computer programmer
Computer scientist
Computer systems analyst

Economist
Engineer
Financial analyst
Information scientist
Life scientist
Mathematician
Operations research analyst
Physical scientist
Social scientist
Statistician
Systems analyst
Systems engineer

Professional Associations

The following lists show some of the associations that relate to the professions of actuary, mathematician, statistician, and operations research analyst. Review the Members/Purpose section for each organization and decide whether the organization is related to your interests.

For additional information, refer to the *Encyclopedia of Associations* published by Gale Research, Inc. Membership in one of these organizations will be useful in providing job listings, networking opportunities, and employment search services. Some provide information at no charge, but others require membership. The same is true for website information. If membership is required, check for student rates. Use the links to other sites provided on these organizations' websites for further career information related to actuaries, mathematicians, statisticians, and operations research analysts.

American Academy of Actuaries
1100 Seventeenth St. NW, Seventh Floor
Washington, DC 20036
actuary.org
Members/Purpose: Qualified actuaries. Seeks to facilitate relations between actuaries and government bodies; conducts public relations activities; promulgates standards of practice for the actuarial profession.
Journals/Publications: *Actuarial Update, American Academy of Actuaries Yearbook, Contingencies, Directory of Actuarial Memberships, Enrolled Actuaries Report*

American Mathematical Society
201 Charles St.
Providence, RI 02904-2294
ams.org
Members/Purpose: Professional society of mathematicians that promotes
  the interests of mathematical scholarship and research; holds institutes,
  short courses, and symposia to further mathematical research; compiles
  statistics; maintains biographical archives; offers placement services;
  compiles statistics.
Journals/Publications: MathSciNet database covering the world's math
  literature since 1940, abstracts of papers presented to the AMS,
  assistantships and fellowships in the mathematical sciences, *Bulletin of
  the AMS*, combined mathematical list, current mathematical
  publications, employment information in the mathematical sciences,
  *Journal of the American Mathematical Society, Leningrad Mathematical
  Journal*, mathematical reviews, more than 496,679 original articles on
  careers and employment links on website

American Society of Pension Professionals and Actuaries
4245 N. Fairfax Dr., Suite 750
Arlington, VA 22203
asppa.org
Members/Purpose: Individuals involved in consulting, administrative, and
  design aspects of the employee benefit business; promotes high standards
  in the profession; provides nine-part educational program; provides
  Certified Pension Consultant Program and administers examinations.
Journals/Publications: *American Society of Pension Actuaries—Yearbook*,
  newsletter, list of courses

American Statistical Association
1429 Duke St.
Alexandria, VA 22314-3415
amstat.org
Members/Purpose: Professional society of persons interested in the theory,
  methodology, and application of statistics to all fields of human
  endeavor. Sections: Bayesian Statistical Science; Biometrics;
  Biopharmaceutical; Business and Economic Statistics; Education;
  Epidemiology; Government Statistics; Physical and Engineering
  Sciences; Quality and Productivity; Social Statistics; Statistical

Computing; Statistical Consulting Education; Statistical Graphics; Statistics and the Environment; Statistics in Marketing; Survey Research Methods; Teaching of Statistics in the Health Sciences.

Journals/Publications: *American Statistician, AMSTAT News,* directory of statisticians; current index to statistics, *Journal of Business and Economic Statistics, Journal of Computational and Graphical Statistics, Journal of Educational Statistics,* proceedings, *Technometrics, STATS*

Association for Symbolic Logic
Box 742 Vassar College
124 Raymond Ave.
Poughkeepsie, NY 12604
aslonline.org

Members/Purpose: Professional society of mathematicians, computer scientists, linguists, and philosophers interested in formal or mathematical logic and related fields. Promotes research in symbolic logic and provides for the exchange of ideas within the international mathematical community.

Journals/Publications: *Journal of Symbolic Logic*

Association for Women in Mathematics
11240 Waples Mill Rd., Suite 200
Fairfax, Virginia 22030
awm-math.org

Members/Purpose: Mathematicians employed by universities, government, and private industry; students. Seeks to improve the status of women in the mathematical profession and make students aware of opportunities for women in the field. Membership is open to all individuals regardless of gender.

Journals/Publications: newsletter: *Association for Women in Mathematics, Directory of Women in the Mathematical Sciences, Careers for Women in Mathematics, Careers That Count, Careers in Mathematics, Profiles of Women in Mathematics: The Emmy Noether Lecturers*

Casualty Actuarial Society
1100 N. Glebe Rd., Suite 600
Arlington, VA 22201
casact.org

Members/Purpose: Professional society of insurance actuaries promoting actuarial and statistical science as applied to insurance problems other than life insurance (such as casualty, fire, and social). Examination required for membership. Provides continuing education and scholarship program.

Journals/Publications: Proceedings, yearbook, also publishes research papers and study notes on actuarial topics, books, journals, online courses

Caucus for Women in Statistics
c/o Cynthia Struthers
200 University Ave. West
Waterloo, ON N23G1
Canada
statwomen.org

Members/Purpose: Primarily statisticians united to improve employment and professional opportunities for women in statistics. Conducts technical sessions concerning statistical studies related to women.

Journals/Publications: Caucus for Women in Statistics directory, newsletter: *Caucus for Women in Statistics*

Conference Board of the Mathematical Sciences
1529 Eighteenth St. NW
Washington, DC 20036
cbmsweb.org

Members/Purpose: An umbrella organization consisting of fourteen professional societies, all of which have a primary objective to increase the knowledge in one or more of the mathematical sciences. Its purpose is to promote understanding and cooperation among these national organizations so that they work together and support each other in their efforts to promote research, improve education, and expand the uses of mathematics. The fourteen professional societies include AMATYC (American Mathematical Association of Two-Year Colleges), AMS (American Mathematical Society), ASA (American Statistical Association), ASL (Association for Symbolic Logic), AWM (Association for Women in Mathematics), ASSM (Association of State Supervisors of Mathematics), BBA (Benjamin Banneker Association), INFORMS (Institute for Operations Research and the Management Sciences), IMS (Institute of Mathematical Statistics), MAA (Mathematical Association of America), NAM (National Association of Mathematicians), NCSM (National Council of Supervisors of Mathematics), NCTM (National Council of Teachers of Mathematics), SIAM

(Society for Industrial and Applied Mathematics), and SOA (Society of Actuaries).

Journals/Publications: Online surveys, reports

Convention/Meeting: Semiannual council meeting

Conference of Consulting Actuaries

1110 W. Lake Cook Rd., Suite 235
Buffalo Grove, IL 60089-1968
ccactuaries.com

Members/Purpose: Full-time consulting actuaries or government actuaries; provides services in the life, health, casualty, and pension fields. Mission is to advance the practice of actuarial consulting by serving the professional needs of actuaries. Annual contest and awards, seminars.

Journals/Publications: *Consulting Actuary*, proceedings of the Conference of Consulting Actuaries

Econometric Society

Northwestern University
Department of Economics
2003 Sheridan Rd.
Evanston, IL 60208-2600
econometricsociety.org

Members/Purpose: Statisticians, mathematicians, and economists. Promotes studies that are directed toward unification of the theoretical and empirical approaches to economic problems and advancement of economic theory in its relation to statistics and mathematics.

Journals/Publications: *Econometrica Journal*, papers, monographs

Institute of Mathematical Statistics

Business Office
P.O. Box 22718
Beachwood, OH 44122
imstat.org

Members/Purpose: Professional society of mathematicians and others interested in mathematical statistics and probability theory.

Journals/Publications: *Annals of Applied Probability, Annals of Probability, Annals of Statistics, Institute of Mathematical Statistics Bulletin, Statistical Science*, probability surveys, electronic journal, books

**Institute for Operations Research and the
Management Sciences**
7240 Parkway Dr., Suite 310
Hanover, MD 21076
informs.org
Members/Purpose: Scientists, educators, and practitioners engaged or
interested in methodological subjects such as optimization, probabilistic
models, decision analysis, and game theory. Also involved in areas of
public concern such as health, energy, urban issues, and defense systems
through industrial applications, including marketing, operations
management, finance, and decision support systems. Operates a visiting
lecturers program; sponsors competitions; offers placement service;
compiles statistics.
Journals/Publications: *Mathematics of Operations Research, Interfaces,
Marketing Science, Operations Research, OR/MS Today, Operations
Research Letters, ORSA Journal on Computing, ORSA/TIMS Bulletin,*
RRSA/TIMS membership directory, *Stochastic Models, Transportation
Science, INFORMS Online*

Insurance Accounting and Systems Association
4705 University Dr., Suite 280
P.O. Box 51340
Durham, NC 27717-3409
iasa.org
Members/Purpose: Insurance companies writing all lines of insurance.
Associate members are statisticians, statistical organizations, actuarial
consultants, independent public accountants, management consultants,
and other organizations related to the insurance industry that are not
specifically eligible for active membership. Educational organization.
Journals/Publications: *The Interpreter,* proceedings, *Life Accounting*
textbook, *Property and Liability Accounting* textbook, CFO/CIO state
report

International Association for Mathematical Geology
4 Cataraqui St., Suite 310
Kingston, ON K7K 127
Canada
iamg.org

Members/Purpose: Professional geologists, mathematicians, statisticians, and interested individuals. Promotes cooperation in the application and use of mathematics and statistics in geological research and technology.

Journals/Publications: *Mathematical Geology, Computers and Geosciences*, IAMG newsletter, *Studies in Mathematical Geology*, newsletters, monographs, research index

Mathematical Association of America
1529 Eighteenth St. NW
Washington, DC 20036-1358
maa.org

Members/Purpose: College mathematics teachers; individuals using mathematics as a tool in a business profession or another profession.

Journals/Publications: *MAA: Online, American Mathematical Monthly, College Mathematics Journal, Mathematical Association of America, Mathematics* magazine

Math/Science Network
Mills College
5000 MacArthur Blvd.
Oakland, CA 94613-1301
expandingyourhorizons.org

Members/Purpose: Mathematicians, scientists, counselors, parents, community leaders, and representatives from business and industry who are interested in increasing the participation of girls and women in mathematics, science, and technology.

Journals/Publications: *Broadcast/Beyond Equals: To Promote the Participation of Women in Mathematics, Expanding Your Horizons in Science and Math: A Handbook for Conference Planners*, science and math resources

National Association of Mathematicians
c/o Dr. Robert Bozeman, Secretary-Treasurer
Department of Mathematics
Morehouse College
Atlanta, GA 30314
math.buffalo.edu/mad/NAM

Members/Purpose: To promote excellence in the mathematical sciences and mathematical development of underrepresented American minorities; conducts annual National Meeting in January; supports an annual invited address at the Joint Mathematical Summer Meetings;

sponsors annual faculty conference on research and teaching excellence; sponsors a Summer Institute in Computational Science for undergraduate mathematics majors and selected faculty.

Journals/Publications: newsletter, job openings

Society of Actuaries

475 N. Martingale, #600

Schaumburg, IL 60173

soa.org

Members/Purpose: Professional organization of 17,000 members trained in the application of mathematical probabilities to the design of insurance, pension, and employee benefit programs. Nonprofit, educational. Sponsors series of examinations leading to designation of Fellow or Associate in the society; maintains 3,200-volume library on actuarial science and statistics.

Journals/Publications: *The Actuary, Journal,* Monograph series, *Directory of Actuarial Memberships,* Society of Actuaries—record, Society of Actuaries—transactions, Society of Actuaries yearbook, index to publications, online member directory, online library. In addition, Society of Actuaries and Casualty Actuarial Society jointly sponsor a website, prepared especially for those interested in an actuarial career: beanactuary.org.

Society for Industrial and Applied Mathematics

3600 University City Science Center

Philadelphia, PA 19104

siam.org

Members/Purpose: To promote research in applied mathematics and computational science; further the application of mathematics to new methods and techniques useful in industry and science; provide for the exchange of information among the mathematical, industrial, and scientific communities.

Journals/Publications: CBMS-NSF Regional Conference Series in Applied Mathematics, *Classics in Applied Mathematics and Proceedings, Frontiers in Applied Mathematics, SIAM Journal on Applied Mathematics, SIAM Journal on Computing, SIAM Journal on Control and Optimization, SIAM Journal on Discrete Mathematics, SIAM Journal on Mathematical Analysis, SIAM Journal on Matrix Analysis and Applications, SIAM Journal on Numerical Analysis, SIAM Journal on Scientific Computing, SIAM News, SIAM Review, Studies in Applied Mathematics, Theory of Probability and Its Applications*

Special Interest Group on Numerical Mathematics
c/o G. W. "Pete" Stewart
University of Maryland
Department of Computer Science
Building 115, A. V. Williams Building
College Park, MD 20742
Members/Purpose: A special interest group of the Association for
Computing Machinery. Individuals interested in computing
mathematics. Encourages communication between members and with
other professional organizations; arranges representation on international
standards committees.
Journals/Publications: newsletter: *The Constant Society*

Young Mathematicians Network
College of Arts and Sciences
Department of Mathematics
University of Kentucky
715 Patterson Office Tower
Lexington, KY 40506-0027
Members/Purpose: A mathematicians' group keeping the mathematical
community honest about the job market and its future; provides
information about job searches from both the inside and outside; is a
support group for those on the job market; provides information on
publishing, grant proposals, obtaining industry jobs, and other things
that many didn't get in graduate school; informs the mathematical
community of the interests and concerns of younger mathematicians.
Journals/Publications: *Concerns for Young Mathematicians*

8

Working Toward an Advanced Degree

Jobs in **marketing, research,** and **financial analysis** frequently require advanced degrees in mathematics or in a related professional field, such as business administration, statistics, computer science, or finance. In addition, an advanced degree is essential for promotion in most careers in these fields. While the job market is expected to be strong in these areas, it is still true that the minimum requirement in many private sector jobs is a master's degree, and that excellent quantitative skills are essential.

Careful thought and planning will be required to accomplish your goal, make the best choices for the future you envision for yourself, and provide for the time and expense that are involved. It is well worth exploring every avenue for scholarships, awards, corporate support, and government and other grants. This chapter is about making your decision to get a master's or a doctorate in the field that you choose—not only to enable you to get the most out of the degree, but to enhance your job prospects.

Considering Your Present Position

Consider the following scenarios to see whether any of them fit your current thinking or give you new ideas about your particular situation.

Scenario 1: You've enjoyed your math major and done pretty well academically, but you're not certain about what kind of work you should look for. You majored in math because you were good at it but really haven't given a lot of thought to how you'll use it. You're confused about what you can do in the workforce and who will hire you.

Scenario 2: You're approaching graduation or you've just graduated. You've tested the job market, and it doesn't seem like a very good time to become employed. Jobs are hard to come by and low paying, and employers don't seem to value your degree enough. You think maybe it would be better to stay in school, get an advanced degree, and then try the job market again, with better credentials.

Scenario 3: Perhaps you've done the kinds of job search activities suggested in the opening chapters of this book and have seen all the better jobs and bigger salaries going to those who have master's degrees. So you think you'll get a master's, but you're worried about your lack of work experience in a professional environment. Most of the math jobs to which you are attracted require a master's degree. You wonder if you could compete successfully for one of these jobs even when you get the degree. You have been wondering whether you have what it takes to compete.

Wrestling over the question of whether it is best to continue your education to get a master's degree or leave school and go to work could have you feeling worried and confused. Given the current uncertainties of the job market and the shifting paradigms of employers and employee bases, it might seem like a wise investment to continue to add to your math education, earn an advanced degree, and hope to make yourself more competitive than the typical undergraduate job candidate.

For those who can afford it, graduate school has always been a popular option when the job market is tight. The strategy to enroll in grad school, wait a couple of years, and enter the job market at a more propitious time is based on the assumption that not only will your chances of employment be better but once you have an advanced degree you may be a candidate for more and different jobs than you would have been previously, and probably at a higher salary.

Definition of the Career Path

Many options are open to you as a math major when it comes to graduate school. If your undergraduate degree is in math education and you are pursuing a career in teaching, then you might consider a master of arts in teaching or a master of education.

Choosing a master's degree in mathematics or a specialty within mathematics would seem obvious, especially for some of the math positions discussed in the previous chapters. However, you have many other choices with your undergraduate mathematics degree. For example, you may be interested in using your undergraduate mathematics degree in a business field, so you might want to consider a master of business administration (M.B.A.) or a master's degree in economics (M.A.Ec.). There are many other choices such as master of arts in finance (M.A.F.), master of applied mathematical science (M.A.M.S.), or a master's in applied statistics (M.Ap.Stat.). Here's a list of a few others to get you thinking:

- Master of Arts in Management (M.A. in Mgt.)
- Master of Arts in Teaching Mathematics (M.A.T.M.)
- Master of Science in Biomathematics (M.S. in Biomath.)
- Master of Computer Science (M.C.S.)
- Master of Finance and Banking (M.F.B.)
- Master of Information Systems (M.I.S.)
- Master of Science in Mathematics (M.S. in Math.)
- Master of Science in Probability and Statistics (M.S. in Probability and Statistics)
- Master of Science in Operations Research (M.S.O.R.)
- Master of Science in Statistics (M.S. in Statistics)

Considerations in Choosing a School

Not all master's degrees are created equal. Graduate programs, curricula, faculty, and financial costs vary considerably from school to school. A master's degree has been referred to as the "working professional degree," because so many management and leadership positions demand an advanced education.

Every graduate school program comes with its own particular faculty, and that faculty has particular professional interests. You'll want to be certain that the research interests of the faculty match your needs in pursuing graduate education. Many graduate students may attend graduate school at night and on weekends as part of the educational benefits offered by an employer. You'll be working two jobs, in effect—your day job and your education after hours. When you consider the energy demands this will require, you will want to be assured that you are getting the kind of education that will serve you best.

A Question of Experience

An advanced degree can, unquestionably, bring more responsibility and greater rewards—especially financial—in the job market. Such rewards are not based entirely on the degree, of course, but also on the quality of experience you can bring to an employer. If you stay on for graduate school and earn your master's degree immediately, you may find yourself in the position of being strong in education but weak in practical experience.

Many positions that require an advanced degree such as a master's include responsibilities like budget management, staff supervision, or strategic planning. Many potential employers will tend to favor applicants who have both a master's degree and the work experience to go with it.

There are exceptions to this rule, though. In certain research or staff positions, your educational experience is of primary importance. Academic success in your particular specialization may qualify you for such a position. If your academic career accomplishments place you in the running for such jobs, your professors and staff advisers will help you plan your advanced degrees and preparation accordingly.

Another important consideration against moving on to graduate work immediately may apply to those who have already acquired good work experience through high-level internships, teaching assistantships, or research assistant positions while undergraduates. Some undergraduate résumés are already rich in high-quality work experiences. If these students can get a good job, with enough income, potential for career development, and ongoing quality of exposure to information in the field, these students may be best advised to enter the workforce and further develop their careers for two to three years more, before deciding on an advanced degree. With the added experience, they may be more focused on their preferred career directions and better able to select the best emphasis and schools for their graduate work.

In many cases, a master's degree without professional work experience means a job in which you're given work that requires your advanced degree but in which your pay level and authority may be those of employees with only an undergraduate mathematics degree. Often applicants with weak experience records but advanced degrees have no trouble securing interviews and offers. Employers are interested in their potential but are often unwilling to place these applicants in jobs with pay levels at the master's degree level without a comparable level of practical experience.

Finding a job is hard work, whatever shape the job market is in. It means putting yourself on the line; answering tough questions; going out on your own, day after day, facing a lot of rejection and wondering why. For these reasons, a graduate degree starts to look good to many math graduates who feel that getting another degree will add to their résumé and maybe, while they're doing that, the job market will change and things will somehow get easier. It's not a wise strategy to continue to add education without experience. Sadly, many of your peers who have taken master's degrees with no real-world experience will tell you that their salaries and job descriptions differ little from those of other employees with only a bachelor's degree. In our career counseling practice, we have seen this type of client all too frequently.

The best reasons to go to graduate school are to expand your education in your field and develop your expertise through intensified study and interaction with peers of advanced knowledge and skill. You will need to analyze your goals and research the part of the field in which you are most interested and the best schools in that area of specialization. You will need to apply to more than one school, since the competition is intense. Many graduate schools receive thousands of applications for every student they accept.

If you are already attending a school that has an outstanding graduate school in your field, you are fortunate, and that school is probably your best choice for many reasons. If, however, you will be looking primarily at other institutions, you have a good deal of time-consuming work ahead of you.

Gaining Admission to Graduate School

For the candidate who approaches the admissions office of a graduate school, two elements of the application are particularly important: a strong personal essay and an interview (even if not required). Both are good indicators of your readiness for a graduate program. With each, admissions committees will be looking for maturity, drive, and focus.

The personal essay required by your application should clearly enunciate your motivation, goals, and degree of readiness. Examining these issues in preparation for such a personal essay is a valuable experience in and of itself. One excellent guide to writing the essay is *How to Write a Winning Personal Statement for Graduate and Professional School* by Richard J. Stelzer.

Even when it is not required, try to have a personal interview at the school. An interview will allow the admissions staff to question you and provide an opportunity for you to explain your motivation for seeking an advanced degree and discuss your commitment to your education.

Suggested Career Path to a Master's Degree

The ideal job for a math undergraduate contemplating graduate school would be a position of some sophistication that utilizes your math education and provides you with opportunities to learn more, a reasonable salary, and a chance to gain the kinds of experiences that will help you if you choose to go on to graduate school. Such positions do exist. These position titles are market analyst, financial analyst, and research analyst or research associate. They may also be called business analysts, associate consultants, or just associates.

Many of these positions are in business consulting, financial services, banking, or computer services consulting. Whether in banking or consulting, the analyst's role is similar. Analysts provide expert advice that will help their employer or their employer's clients invest wisely to get the best return on their money or solve business problems.

Market and Financial Analysts

Your work will involve evaluating the marketplace as a whole. You'll study information on shifts in gross national production, cost of living, personal income growth, rates of employment, construction starts, fiscal plans of the federal government, growth and inflation rates, balance of payments, market trends, indexes of common stocks, and the like.

You also need to be aware of national and international events that could precipitate some serious reaction in the marketplace. You will monitor both the data on your screen and key social, political, and economic factors that are affecting the pulse of the world. The position of analyst is a fascinating and multifaceted job for a math graduate.

Following is an ad for a market analyst. It's a fairly specific job listing that contains important information about the nature of the work and the qualifications being sought.

Market/Financial Analyst—Entry Level

XYZ Analysis, Inc., is a management consulting company that applies sophisticated analytical techniques to business problems in the public and private sectors. XYZ's strength is in high-quality, state-of-the-art services in the areas of strategic planning, decision analysis, operations management, public policy analysis, market forecasting, research and product development planning, and basic research. Recent clients include General Motors, Xerox, Kaiser Permanente, the Electric Power Research Institute, the U.S. Department of Energy, and the U.S. Environmental Protection Agency.

XYZ is seeking applicants with bachelor's degrees for positions in our analytical staff. Candidates should have: four-year degree in mathematics, operations research, decision analysis, computer science, engineering, or other technical field; a GPA of 3.5 or higher; interest in solving complex problems; skills in a broad range of mathematical techniques; communication skills to present analytical results in a clear, concise manner; professional and Microsoft Office computer expertise, including Word, Excel, and PowerPoint; ability to prepare and present effective client deliverables including reports and presentations; ability to work at high productivity levels in team structures; and high level of enthusiasm for challenging, fast-paced schedules.

Analysts at XYZ work on teams with other XYZ consultants on a variety of projects. Analysts are involved in tasks such as data analysis, formulating and programming mathematical models, working with clients, and preparing, presenting, and assisting with report and proposal writing.

Several things are clear in this ad. There are high expectations for your performance, but the rewards are commensurate. As we've discussed throughout this book and this ad corroborates, you need to know your mathematics, and this employer specifies a high GPA to ensure that. At the same time, there is a clear indication that working conditions are relaxed and interesting.

Note the emphasis on communication skills and the fact that you'll be involved in client proposals, multimedia presentations, and a number of writing assignments as well. If your math courses have not demanded much writing, you might want to consider a technical writing class in your English department or an organizational communications course in the business department of your college.

Research Analyst and Associate

As a research analyst, you will be doing original, Internet, and library research, collecting data in organized forms, and conducting some data manipulations and analyses. Research associates and research analysts assist in helping put together proposals, case studies, or analyses designed to help the consultant's client solve problems, determine future strategies, or implement programs.

As you gain expertise, some responsibilities will be added, in most cases having to do with additional and more sophisticated research capabilities, quantitative manipulation of data using computer software, and presentation of findings with your work team. However, there is a limit on what you can do, on the decisions you will be allowed to make, and in how far you can go on your own. These positions are structured to work under more senior analysts or other managers. Senior workers will have the major decision-making responsibilities. The positive aspect of this supervision is that you'll work closely with experienced professionals who will have much to teach you.

The following ad is for a research analyst position for someone just out of college:

Research Analyst, Entry Level

XYZ is an innovative world leader in telecommunications and networking, based in the Salt Lake City area, and with offices in India and Mexico. We are a high-tech, high-growth company in a rapidly changing global industry. XYZ is a diverse company, with a more than 80 percent mutlilingual workforce. Our global network provides voice, data, video, Internet access, satellite TV, and messaging. XYZ has entry-level opportunities for individuals who seek employment involved with the development of business systems. Selected

---

continued

> candidates will participate in an intensive six-week training course. At the conclusion of training, candidates will be assigned to one of the company's internal organizations providing business analysis support in the resolution of key market problems. After a successful six-month performance evaluation, candidates will be positioned as intermediate-level analysts.
>
> Key topics in the classroom training are:
> - Problem analysis
> - Function/task analysis
> - Data analysis
> - Systems design/test plan techniques
>
> We require a BA/BS degree in mathematics (with computer option), information systems, or quantitative/systems analysis, with minimum GPA of 3.5 in major.
>
> XYZ is aggressive in attracting the best, with generous salaries, benefits, and growth potential.

This position description shares some similarities with the market research position listed earlier. There's a sense of high expectation for performance and, like the previous ad, this firm is also demanding a specific GPA level. Both firms have obviously decided that grades are a reasonable indicator of ability, proof positive that, for some employers, grades count.

This ad also emphasizes a higher degree of analytical functions compared to the market analyst position and specifically enumerates a number of kinds of analysis that will be part of your training. The emphasis on training is good news. Professional training by employers is often exceptional in its quality and stays with your career beyond any one job. It also is an indicator of stability, as organizations that spend resources training employees have a vested interest in retaining them and seeing them grow and develop.

A final note of importance is that the submission of a résumé and cover letter as e-mail (with submission specifications) or as a scannable document means this firm is operating in the electronic business world and takes it for granted that applicants will be also.

For guidance on e-mail submissions and preparing scannable résumés, stop by your career office and get some assistance. Joyce Lain Kennedy has an excellent book on the market titled *Electronic Resume Revolution* that addresses writing the best scannable résumés as well as using electronic databases to keep yourself visible. Another publication, the *Guide to Internet Job Searching* by Riley and Roehm, published by McGraw-Hill, provides instructions for posting a résumé electronically.

These positions are featured in this chapter because they particularly appeal to graduates contemplating advanced education. There's a fairly steady turnover in these analyst jobs as people choose to return to school or are promoted within the firm. More people move on, however, than move up because an advanced degree is often de rigeur to move up in these kinds of organizations. So, since these positions are not designed to fulfill career aspirations in and of themselves, they should appeal to you if you're contemplating graduate school.

Though they come with excellent salary and benefits, they have more in common with postgraduate internships than typical entry-level positions, because they typically involve so much education and training. In fact, these analyst positions are often part of special hiring "programs" for recent college graduates. Since the jobs do generally turn over rather quickly, many entry-level candidates are recruited and hired.

The training and mentoring involved for the new analysts are considerable. The responsibilities for training generally rest with one individual or are centered in an office that manages the research associate/analyst program. The defined boundaries of the experience and the structure of the internal training process should be highly acceptable to the mathematics student with a bachelor's degree seeking to gain valuable skills and experience before entering graduate school.

Strategic Advantages

Job candidates applying for these positions do not have to misrepresent their intentions of eventually going on for a graduate degree. In fact, if you should desire to stay, you would find on investigation that senior responsibility and authority are reserved for the higher-degreed specialists.

Many analysts do, in fact, continue to work for, or return to, the consulting firms and banks that initially hired them after obtaining their master's in any of a number of majors, such as quantitative analysis or business administration. At this point, they begin new career paths, in positions with titles such as associate-to-consultant, investment manager, or perhaps chartered financial analyst (a designation similar to CPA received after passing equally rigorous examinations).

After some time spent as a research associate, you will have significant, important business experience that will have transformed the value of your undergraduate degree. You will be better situated to apply to a good graduate program and feel confident about your ability to succeed. Whatever graduate program you enter, you will arrive better prepared to make the most of the degree. When you do graduate with your new degree, you will offer your new employer excellent justification for a responsible, decision-making position with all the appropriate rewards of such a job.

An Alternative Strategy

Perhaps in reading through this strategy, you've begun to have some doubts about its suitability in your case. Though you agree with the proposition, you don't feel like the ideal candidate for an advanced degree. Your objections might be on any number of grounds: your grades in college may not have been good enough for you to be competitive for some of these analyst positions, or, perhaps, you are not interested in living for any amount of time in a metropolitan area where these businesses are often located. It may be that the nature of the work of a research associate does not interest you. Some folks just get tired of school and have no wish to return immediately.

There are also other options. The basic premise of this book is to give a math major frank advice about the worth of this degree and how best to use it in forging a productive and enjoyable career. One way you can ensure that outcome without going to graduate school is by focusing on the development and acquisition of portable skills that you can carry from job to job rather than the contextual skills that are pertinent to one job and not easily transferable to any other.

If you work in the investment department of a bank and monitor the bank's portfolio, you acquire two kinds of knowledge in this job. The first, your content knowledge, is what you learn about those specific stocks and their performance. Unless you should happen to encounter those stocks again in another job, the information does not "carry over" to new employment. Portable skills, however, follow you throughout your career. The portable skills you learned in portfolio management would be data analysis, work with relational databases, and knowledge of the market and trends in general. In today and tomorrow's ever-changing job market, portable skills offer you the best measure of job security.

If the path to a graduate degree is not for you, you can still accomplish the same goals in a slightly different way. You may seek out positions after graduation that provide the same kind of preparation as the analyst jobs can give you but without some of the "thorns." You may be able to find similar positions without relocating to a densely urban area; you can locate analyst positions in firms with less turnover and less competition for the entry-level jobs. You certainly can locate firms that will value your talents and want to keep you.

You may decide to go to graduate school at night and hold on to your position. An employer may offer you educational benefits as an inducement.

You might also decide your job is so interesting with so many possibilities for personal enrichment that you decide not to go to graduate school, at least for the foreseeable future. Whatever your decision, you will have begun to build on your undergraduate education with valuable work experience that benefits both your career and future professional education.

Working Conditions

The best analysts read constantly—newspapers (more than one a day), annual reports, trade publications, journals, biographies, even novels—whatever keeps them abreast of the developments and changes in the marketplace. This may come as a surprise, delight, or concern to you, because much of the work in your major involved processing and problem solving and not a significant amount of reading. In addition, college life often leaves busy undergraduates little time to stay abreast of current affairs. If you are a reader and enjoy staying current, then you'll be delighted to find a career path that combines that interest with mathematics. It will perhaps be a concern if you are not only out of practice in general reading but are not a particularly avid reader when you do have time.

Geographic Concentration of Potential Employers

These jobs are not available everywhere. Whether you are working for a consulting firm of the size that can afford a number of research or financial associate positions, or an investment bank, or the investment department of a commercial bank large enough to have such a department, you should be thinking in terms of a location in a major metropolitan area or the highway belt around such an area. This is where you will find the greatest concentration of such businesses. Even more specifically, most of these positions are concentrated in the major metropolitan areas of Boston, Chicago, New York City, and San Francisco. The major brokerage houses, however, also have branch offices in more than eight hundred cities across the United States.

Living in a Metropolitan Area

If you have not lived in a major metropolitan area, you should explore the myriad lifestyles that are available. There are a variety of reactions to living in a metropolitan area—some people love it and others don't. The cost of living is higher than in a smaller city, and certain tensions caused by the fast pace of living and security issues exist as well. You will need to prepare yourself with as much advance information and familiarity with the new environment as possible. You can benefit from reading a straightforward tourist's guidebook to the city before you accept a job or invest in a move.

Most people will admit that, in spite of many difficulties involved, metropolitan living offers more choices than any other locale: in jobs, culture, shopping, entertainment, and opportunities to meet people. Some have found that living in a city neighborhood can resemble living in a small-town neighborhood. They see the same people on their way to and from work, shopkeepers who know their names, waitstaffs who know their breakfast and lunch orders, and so forth. Many city dwellers are quick to describe the advantages and the "normalcy" of city life. Many fine graduate schools are also located in large population centers as well, and that could prove to be convenient when it comes time to move ahead in your strategic plan to enter graduate school.

What's It Like to Work in Consulting?

There is no "typical" day for a research associate or financial analyst in consulting or investment banking. However, while there may be no rigid daily routines (which may be an attraction for you), the following activities are fairly constant over time.

Information Resourcing

Associate positions require finding answers, usually under pressure of time and cost. To succeed, you need to be inventive and good on the phone, and believe that you can do it to succeed. Most consulting and investment decisions require you to have high-quality, timely information. Providing that information will be a big part of your daily job. Your day will probably begin by scanning several newspapers, watching for business and market information that might prove helpful. Begin that reading habit now. It'll come in very handy.

Financial analysts at investment banks or in the investment department of larger banks monitor the performance of particular stocks and securities on the world's stock exchanges and stay well informed about the industries they track. Generally, an analyst will focus on one or two industries in this kind of position.

Travel

Long-distance travel is more a requirement for a research associate in the consulting industry than for an analyst in investment banking. Some entry-level consulting positions can involve a grueling amount of travel. Just as the problems are global, clients can be global as well.

Consultants leave the United States every day for Africa, China, Eastern Europe, India, Korea, Russia, and South America. Many large consulting firms and investment banking services have had a strong European presence for decades. This increasingly far-flung demand for consulting/analyst services gives an edge to the person with language skill and/or cultural sensitivity.

Both research and financial analyst positions can involve significant out-of-office time in meeting with on-site clients. To succeed, you must be prepared and willing to go, at very short notice. Projects may require you to be away from home for several weeks or only a day. When you are on-site with a client, away from home, workdays will usually be long, because your client wants to make the most of your time with them and also knows that you are not going home at the end of the day but to a hotel. Long workdays, with working dinners, as well as working breakfasts the next morning, are not unusual.

Statistical Analysis

Much of your time as a statistical analyst will be spent designing and/or running complicated computer models to evaluate corporations, monitoring stock price movements, or working with numerous databases to retrieve important information. You will need to become familiar with all kinds of statistical digests, annual reports, Securities and Exchange Commission documents, statistical and financial software, spreadsheets, and financial market reports.

Whether you need to analyze the investment potential of a foreign firm, the comparable company activity pursuant to a merger, or the involved preparations prior to a public offering of stock, you will encounter data that need to be transformed into usable information by your analysis. This information will then become the basis for the consulting team's action plan for the client.

Professional-level skills in the manipulation and understanding of data will provide some of the most valuable experience you can bring to your future graduate program. It is perhaps one of the most crucial skills for graduate work; and your ability to transform data into information should help make your progress through graduate school a smooth one. You will have an advantage over many of your peers, even those with business experience, because as a consulting associate or financial analyst, you are working for so many different clients on so many different projects that you will develop a broad understanding of resources and relevant techniques.

Presentations of Client Deliverables

The impact of PowerPoint presentations in the business world today is nothing short of phenomenal. In a consulting firm, for example, the work is underwritten by client fees. Though you may be working on several different projects simultaneously, all your work is client directed and available to the client for examination and review. Attractive presentation of materials is a skill you will begin to appreciate in consulting. The presentation may be a carefully prepared written document with charts and graphs (demanding mastery of a variety of software products), or it may be a public presentation with overheads, handouts, and explanations of your material, as well as responses to questions. Highly developed creativity and communication skills, quality work within tight deadlines, and skills with a variety of state-of-the-art multimedia presentation techniques are some of the professional expertise that you will take away from this experience.

In some companies, the analyst is a one-person multimedia department, and in others the client deliverables constitute a full publishing operation. Large corporations may have a publications or production department that will assist you in preparing proposals and client deliverables. Writers, editors, graphic designers, and coordinators may work with you on major projects.

It is not unusual in companies like these for an in-house printing and binding department to take over the last parts of the project, and complete dozens or hundreds of printed and bound analyses, reports, and recommendations to be delivered to the client. These operations are often done under very tight deadlines, and you may receive additional information from a client or an internal manager that must be analyzed, written, and incorporated—overnight.

Teamwork

As a research associate or analyst, your work supports more senior staff people who have the major responsibility for the success or failure of a client contract. Consequently, a team approach is essential to every project you undertake. You'll learn to take the responsibility of communicating your activities promptly and clearly to other team members in order to avoid overlaps, misunderstandings, and wasted effort on everyone's part. You will learn to evaluate your coworkers' levels of cooperation and follow-through, as well as your own, and find ways to ensure that the level of accomplishment lives up to your employer's and the clients' needs and expectations. This intense experience in a team approach may surprise you, especially in how it adds to your level of maturity; and it will be an asset in your career, whether you decide to continue in the working world or pursue another degree in graduate work.

Training and Qualifications

Requirements and opportunities for training vary greatly. Let's begin with a general statement on training for research assistants and financial analysts from the *Occupational Outlook Handbook*:

> There are no universal educational requirements for entry-level jobs in this field. However, employers in private industry prefer to hire those with a master's degree in business administration or a discipline related to the firm's area of specialization. Those individuals hired straight out of school with only a bachelor's degree are likely to work as research associates or junior consultants, rather than full-fledged management consultants. It is possible for research associates to advance up the career ladder if they demonstrate a strong aptitude for consulting, but, more often, they need to get an advanced degree to do so.

Employers of these positions are looking for people with very specific qualifications. A qualified candidate will possess strong quantitative, analytical, and communication skills. He or she will be able to work in a fast-paced, demanding environment and will need to demonstrate achievements such as a high GPA, sophisticated academic and part-time work experiences, and extracurricular leadership positions.

Financial analysis and research are done in a wide variety of contexts by some of the most prestigious firms in the world. Generally, what these firms look for in applicants, both on their résumés and in personal interviews, is evidence of facility in mathematics; the ability to digest, analyze, and interpret large amounts of material; an inquiring mind; and good communication skills.

Listening is also a critical skill. When you are dealing with clients, unless you understand their problem as they do, you are wasting time and money in devising less-than-sufficient solutions. Consulting is about communication. The following checklist presents some of the essential skills and their level of importance in consulting:

- Ability to synthesize: *high*
- Analytical skills: *extremely high*
- Communication skills: *high*
- Computer skills: *medium*
- Creative ability: *high*
- Initiative: *medium*
- People skills: *high*
- Sales skills: *medium*

Although your background may be in mathematics, you will be joined in this career area by colleagues with degrees in economics, political science, business administration, finance, marketing, engineering, and law. Any of these degrees, often combined with an advanced degree in business administration, is considered very attractive by hiring firms.

Earnings

Salaries in this field vary dramatically, sometimes almost astronomically, depending on the new employee's education, experience, and employer. Median income for analysts is about $53,810. The middle 50 percent earned between $38,760 and $76,310. In your own search for analysts' salaries, you'll need to be watchful. Frequently on the Web, the term *analyst* will refer to data analyst positions, not the analyst positions we've described in this chapter.

What you need to remember in reading these very generous salary figures is that these firms continually have their employees under tremendous scrutiny. If you don't perform, you'll be let go. This remains true even as you ascend the earnings ladder. Most of these firms have specific expectations for their employees' levels of performance. When you see high earnings such as these, you should not be surprised that the following are true: (1) getting hired will be difficult, and (2) you will work very hard indeed for that level of salary.

Career Outlook

Extreme financial pressures in the volatile business environment following the boom of the 1990s have challenged even smaller corporations with issues of global competition, dramatic technological advances, and the need for an endless stream of new thinking and practices for competitive hiring and firing and efficient organizational structure. Outsourcing, whether from domestic corporations or companies in other countries, has been used as a cost-cutting strategy by a majority of large firms. Mergers, acquisitions, joint and allied partnership arrangements, venture capital investing, and other strategies have washed over the corporate scene like a tidal wave in recent years. Faced with this kind of dizzying environment, it's no wonder management cries, "Get us an analyst!"

Employment projections for analysts and consultants are for faster-than-average growth through the year 2012, as industry and government increasingly outsource these jobs to improve their performance. Companies today are more likely to "buy" from among this growing array of specific consulting services to solve problems rather than hire permanent staff to accomplish the same things. It's cheaper and more efficient. As a result of this outsourcing practice, however, growth is expected to occur in large consulting firms, large financial services organizations, and the largest banks.

In today's competitive marketplace, consultants, banks, and financial service organizations offer many similar services to their clients. Some clients may be in trouble over product development, staffing issues, profit and loss projections, or any of a host of complicated situations. Others are doing well and need assistance coping with rapid growth; still others need help increasing profitability or efficiency. Consulting can be utilized by any business for any aspect of its operation.

Strategies for Finding Jobs

Our strong recommendation is obviously to get work experience before you go for an advanced degree. The career path to a possible master's degree that we discussed in this chapter is structured around the strategy of first finding a position as an analyst. These analyst positions go by a variety of names, including market and financial analyst, research analyst, and research associate.

Though these analyst and research positions are not all the same job, they share an emphasis in hiring the mathematics graduate. For these positions, your math education is essential. To find both the advertised and unadvertised analyst positions, begin by identifying your particular skills and then research the jobs, identify the industries, prepare your tools, and begin the networking process.

You will want to locate the analyst positions that match your skills and experience. Let's look at the skills employers are looking for in these positions:

- Exceptionally strong mathematics background
- High GPA
- Presentation skills
- Good oral and written communication
- Teamwork
- Good telephone skills
- Inquiring mind
- Research ability for business and market trends
- Good comfort level with manipulation, analysis, and interpretation of large amounts of data
- Potential ability to analyze financial market reports

From the beginning of your search, look for positions that will value your undergraduate degree and offer training or possible benefits for continuing your education.

Here are excerpts from two typical job postings for an analyst position:

Candidates are required to have an undergraduate degree with a distinguished record of leadership and academic achievement. Additional qualifications include excellent quantitative skills, ability to communicate effectively, and genuine dedication to teamwork, as well as high energy, personal integrity, initiative, and creativity.

Candidates should have solid understanding of basic mathematical concepts, effective writing and communication skills, excellent computer programming skills, and interest in applying these strengths to real-world problems. Other responsibilities include coordinating meetings, managing fieldwork, and assisting with report and proposal writing.

As you read these ads, you will not be surprised at the need for strong math skills. Many math students are surprised, however, at the emphasis on public presentations, excellence in writing, and client conferencing skills. If you're surprised as well, it's probably because none of your college math classes gave you experience in these areas, nor did they indicate that they could be important in your career. They are, however, very important and will be a strong theme throughout this chapter. If you have had experience in preparing and using visual and multimedia presentations, as well as Internet presentations, you will find this experience is an advantage in the job market and will be highly useful in the job itself.

Locate the Right Potential Jobs

Exploring the concentration of jobs that are found in major metropolitan areas or the highway belts around such areas as Boston, Chicago, New York City, and San Francisco, as we have mentioned earlier, is a first step. The kinds of employers that use analysts frequently are concentrated in densely populated areas, in transportation hubs, and near major airports. A search on the Internet for local jobs, by area, will be an efficient way to start.

Identify the Possible Industries

As we continue our job search strategy, using Boston as the geographic area, let's identify the possible employers. For example, here is a sample of the kinds of employers for analyst positions in the Boston area:

- Accounting
- Advertising and public relations
- Banking/savings and loan
- Business services/nonscientific research
- Charities and social services
- Communications
- Education services
- Financial services
- Government
- Health care: services, equipment, and supplies
- Insurance
- Management consulting

Use this list as a guide to find other industries in the area in which you are interested. Search for listings of marketing research firms. Companies turn to these firms for contracting-out services rather than supporting their own marketing department. Explore financial services, advertising, and those manufacturing firms producing consumer goods. Consulting firms, investment banking, and brokerage houses offer excellent employment prospects for analysts. A variety of government agencies provide the largest number of positions, but don't overlook economic consulting firms or financial institutions.

Prepare Your Job Search Tools

Design your résumé to focus on the skills of the position description that you find in the position posting or ad. Modify your résumé to emphasize the areas the employer is looking for. For example, move those skill descriptions to the front of paragraphs or to the top of any lists that you have included. Use the language and vocabulary of the ad if your terms are slightly different but actually mean the same thing. Be sure to use the most current terms if the ad does so, even if they seem like pretentious buzzwords. If they are the everyday-usage terms of the firm, they have a real meaning in practical application in that company or organization.

Write your cover letter so that it targets the benefits of hiring you and describes your math education, skills, and experience in a way that draws parallels to the needs described in the position listing. If the hiring official finds a match between your skills and the employer's qualifications and needs, there is a good possibility they will ask to interview you. If an interview confirms the match between your skills and their required qualifications, then you will likely be hired.

Begin the Networking Process

Once you have identified your most relevant skills, located the jobs and possible employment sites, and have your job search tools ready to submit, you can begin making telephone calls to your contacts to establish your network. In creating your list of contacts, be sure to use your personal and professional contacts, including your college math professors, other students in your courses, colleagues from clubs on campus, and fellow members of organizations on and off campus. Let everyone know what positions you are searching for and in what geographic area. A section at the end of this chapter lists professional associations that have job listings. Check their websites to see whether they publish job postings in their newsletters or journals as well as online.

Market and Financial Analysts

As we learned earlier in this chapter, most analyst positions are found in metropolitan areas. Analysts work in large organizations whose success depends on a knowledge of the general business and financial climate. Some of the important examples are as follows:

- Major corporations
- Financial institutions
- Major teaching and research hospitals and HMOs
- Brokerage firms
- Investment companies
- Pharmaceutical companies
- Major transportation systems

No matter where analysts work, their role is relatively the same: to provide expert advice that will help their employer or their employer's clients invest wisely to get the best return on their money or anticipate or solve a business problem. Because analyst positions have a variety of job titles, it's best to focus carefully on the details of the job description and not on what the job is called. The following is a sample ad for an analyst found on the Internet:

Marketing Analyst: Major Baltimore bank seeks marketing analyst to gather and manage data on consumer products and customer credit profiles. Perform custom segment analysis, profitability analysis, volume/trend forecasting, and budgeting. Provide recommendations to improve effectiveness of direct mail program. Requires application of statistical methods, financial analysis, and experience with managing and reporting from large databases.

Research Analyst and Associate

Client deliverables are the reports and analyses provided to the clients—either in-house clients such as specific departments or an employer's outside clients—during and at the end of the consulting contract. Research analysts and associates put together proposals, case studies, or analyses designed to help the consultant's client solve problems, determine future strategies, or implement programs. These jobs are often seen as opportunities to gain experience that leads to increased responsibility.

Most of these positions are designed to work either in a team or under collaborative supervision with a more senior team member. These positions appeal to graduates contemplating advanced education, because the experience is applicable to their math background and turnover in these positions is normal and expected as job holders often move on to graduate school or more senior positions in the same or other companies.

As you begin to review positions for research analyst and associate, pay close attention to the skills required. On-the-job training is common in these positions, and you will want to devote some interview time to exploring how new analysts are trained in the company you are interviewing with. Here are two more sample ads for positions that emphasize the skills of data analysis, running mathematical models, and proposal writing. Some of you will have some of the computer experience or software familiarity required, and others will have only limited or partial skills in this area. Don't let that deter you from applying. If your computer skills are solid and you've had some exposure to statistical software and feel you have an ability to quickly learn other programs, express that in your cover letter or during an interview. Given the variety of software packages in use today, exposure to the specific brand is less important than familiarity with the type of software and an ability to read a user's manual and ask questions.

Research Analyst: Seeking applicants with technical BA or BS degrees and strong quantitative orientation for the position of research analyst. Candidates for this job should have a solid understanding of basic mathematical concepts, effective writing and communication skills, good computer programming skills, and an interest in applying these strengths to real-world problems. Research analysts work on teams with other consultants on a variety of projects. Research analysts are primarily responsible for tasks such as data analysis, programming and running mathematical models, and programming computer-aided interviews. Other responsibilities include coordinating meetings, managing fieldwork, and assisting with report and proposal writing.

Financial Analyst: Financial analyst new-hires will experience intensive training module for approximately six weeks during initial stage of program. This phase focuses on corporate credit analysis and risk assessment, oral and written communications, accounting, and fundamental corporate finance skills. Concepts explored through classroom lectures as well as individual and group cases. Subsequent to the training module, each analyst will be assigned to a client team and will analyze clients' current operating performance, financial condition, and industry. In addition, analysts will participate in transactions, including preparation of financial projections and client proposals. Successful candidates will be self-motivated and inquisitive, with strong analytical, as well as excellent oral and written communication skills.

Required: BS/BA degree with solid background in accounting or finance, plus quantitative analysis skills. Must pass security clearance and background check.

The intensive training provided in this kind of position is highly desirable. Many of the skills learned are portable skills, and this training will stay with you throughout your career. Firms who invest that much time and money on your training have every reason to assist you in your career progress with their firm. They want you to succeed. You also need to know that positions providing this much training are highly competitive, the selection process may be lengthy and rigorous, and your application and candidacy will come under close scrutiny.

Possible Job Titles

Job titles will not be a reliable guide to analyst positions, so it is necessary to pay as much attention to the job description as to the position title. When you are searching for possible jobs, you will need to be familiar with the scope of possible titles under which your kind of job might be listed. You should maintain an ongoing list of these titles as you pursue your job search so that you gradually become familiar with the many possible variations. Some of the most commonly used titles are as follows:

Associate
Associate consultant
Associate to consultant
Business analyst
Chartered financial analyst
Financial analyst
Financial assistant
Financial auditor
Financial evaluator (trainee)
Investment manager
Market analyst
Research analyst
Research associate

Related Occupations

Most of the valuable skills demanded by analyst positions are highly portable, and the opportunities for transferring these skills to other occupations are excellent. Naming all of those related occupations would be a challenge because they are so numerous and so diverse.

In this chapter we have attempted to give the math major a good idea of the essential job skills needed as you look for analyst positions on your way to a master's degree. To identify other related occupations, compare the skills required for any particular market or financial analyst, research analyst, or associate. A brief list of the skills that employers are currently listing for these analyst positions includes the following:

Math skills
Presentation skills
Communication skills
Teamwork skills
Telephone skills
Willingness to travel
Inquiring mind
Ability to research business and market trends
Ability to manipulate and understand data
Financial analytical skills

As an example, if your particular employment involved client consulting skills, problem solving, data analysis, and constructing mathematical models, a change of careers might easily lead you to work in pension plan or retirement funding programs. The titles of some standard related occupations follow:

Actuary
Budget officer
Credit analyst
Economist
Financial manager
Loan officer
Underwriter

Professional Associations

A selection of major professional associations related to market and financial analysts, research analysts, and associates follows. Membership in one or more of these organizations might be helpful in terms of job listings, networking opportunities, and employment search services. Some provide information at no charge. In some cases, if you want to receive their publications that list job opportunities, a membership may be required.

Check out each one of the organizations and read its home page to see how its services and membership advantages align with your professional and job search needs. You may also want to check for student membership rates. For additional information, refer to the *Encyclopedia of Associations* published by Gale Research, Inc., which is available in the reference department of most libraries.

American Management Association
1601 Broadway
New York, NY 10019
amanet.org
Members/Purpose: Seeks to broaden members' management knowledge and skills.
Journals/Publications: Free e-newsletters, extensive catalog of books, seminars, e-learning materials

Association of Professional Consultants
P.O. Box 51193
Irvine, CA 92619-1193
consultapc.org
Members/Purpose: Professional consultants. Aids members in the improvement of their professional abilities.
Journals/Publications: Membership directory

Council of American Survey Research Organizations
170 North Country Rd., Suite 4
Port Jefferson, NY 11777
casro.org
Members/Purpose: Survey research companies in the United States. Provides a vehicle whereby survey research companies can interact with one another, sharing relevant information and addressing common problems; promotes the establishment, maintenance, and improvement of professional standards in survey research.
Journals/Publications: Membership directory; *ISO Draft International Standard for Market, Opinion, and Social Research*

Institute of Managing Consultants, USA, Inc.
2025 M St. NW, Suite 800
Washington, DC 20036-3309
imcusa.org
Members/Purpose: Professional management consulting firms that serve all
types of business and industry.
Journals/Publications: Directory of membership, *Journal of Management
Consulting*, newsletter

Marketing Research Association
1344 Silas Deane Hwy., Suite 306
Rocky Hill, CT 06067-1342
mra-net.org
Members/Purpose: Companies and individuals involved in any area of
marketing research such as data collection, research, or as an end user.
Journals/Publications: Membership roster, Annual Blue Book Research
Service directory, field service manual, research guidelines, field directors
manual, interviewer training manual

9

Math in the Marketplace

If you enjoy a dynamic work environment characterized by high energy and fast-paced change, you may be interested in this group of jobs. In this chapter, we present the careers of **buyer, sales representative**, and **purchasing agent**, because these careers are based on a firm foundation in mathematics. People in these careers use math as an important core job skill for competing in the commercial marketplace.

Because the marketplace is always seeking and competing for buyers and sellers, this is a field where decisions are often made quickly and in a highly charged atmosphere of financial risk. Innovation, the ability to seize opportunity, intuitive knowledge of the marketplace and of people, and a quick response mechanism are essential characteristics for success in this environment. If these descriptions are appealing to you, and characteristic of your personality and work style, you may want to consider exploring one of these career paths.

In any of these careers, there is no question that you will be needed and appreciated. In today's competitive marketplace, profit margins can be slim, and costs must be carefully controlled. Business strategists are increasingly dependent on sophisticated mathematical models to anticipate possible outcomes to changes in variables such as the price of raw materials; advertising costs; increasing pressures of competition, often from all over the world; volatility of stock markets; packaging and shipping costs; and a host of other variables that can affect the success of a business. The math graduate's skills fit right into the most pressing needs of organizations that are working under these demands and pressures, and can make a significant contribution that may often make the difference between the success or failure of an entire venture.

Each job brings its own environment and its own "package" of associated values. While most readers of this book may be math majors, each person reading this book is also unique and has specific needs. After reading Chapter 1, "The Self-Assessment," and working through some of those exercises, you will have begun to build a mental picture of your particular personality dimensions, work attributes, and values. From your own past work experience, you are aware that each job is unique in its atmosphere, interactions with others, pay scales, and quality of work experiences, which includes such things as respect for products, services, and corporate values; the corporation's respect for its employees; pace of the work; location of the business; degrees of organization and preparedness; and overall culture and traditions of the company.

For career counselors, one of the most enjoyable aspects of working with math majors is that nearly everyone respects the skills and talents of, and wants to hire, a math major! Frequently, career counselors hear employers and recruiters complain that math majors never interview with them. They wonder aloud, "Where do we find them? How can we get them to interview with us?" Even the faculty of a math department may fail to realize the broad variety of careers in which a math graduate can be employed.

Employers need people with mathematical skills in the workplace. Many math majors are unaware of this demand and may have neared graduation and begun their own, often frustrating, job search before they have any realistic idea of the demand for their highly developed skills.

It may be easy to assign blame for the lack of job-market awareness to your program or faculty, but that is not actually fair. A math major is one of the most demanding scholastic majors offered in a college—so demanding, in fact, that the small numbers of graduates are a good indication of your rarity and value in the marketplace. Your college math curriculum requires a packed schedule that does not leave much time for direct application instruction by the faculty. It is not uncommon for students to be so involved in the demands of their program that it is impossible to allow time for exploring career options. That's understandable, but eventually the research and preparation for the job search and career plan must be undertaken and done well.

The purpose of this book, and especially of this chapter, is to focus on areas you may not yet have encountered as career options. This chapter is dedicated to answering questions that you may have had on your mind for quite a while, such as "What do I need to know about working in the area of buying and selling?" We will provide information on preparation, aptitudes, and skills that the job may demand of you in addition to your math skills. Other areas include, but are not limited to, the following:

- Interest in and awareness of the commercial marketplace
- Appreciation for the competitiveness of the marketplace
- Willingness to assume management responsibilities, including staff supervision, budgets, hiring, and evaluation of employees
- Increasing responsibilities for strategic concerns, planning, evaluation, and strategies

This is a career area in which you are hired for your math degree and will use your math and related skills every day. It is also an area in which you will be expected to assume management responsibility for the preparation of professional staff work that will involve writing business reports, memos, and correspondence; staff supervision of employees; and your ability to represent the organization in a responsible manner. Let's look at some specific occupational areas in this career area.

Definition of the Career Path

Many of the opportunities in this career path involve exciting, but perhaps unfamiliar, opportunities for the math major, so let's begin by identifying some possible occupations in this area and acquaint you with some of the terminology, job definitions, and typical career pathways in those fields. A caution is in order here: the jobs described below are only a few solid examples of the broad variety of jobs in this group and are not a complete list by any means. Later in this chapter, we will provide more ideas of possible job titles and related occupations for this area.

We have selected these jobs because (1) they are in areas that seek math majors, (2) math majors are in demand in the present marketplace, and (3) they are good general examples of career paths and job demands in the areas of buying, selling, and marketing. These jobs value your math education highly but demand other skills and interests such as decision making, customer service, or listening skills. In this category of jobs, math is but one of the many important tools in your skill package you will need to do these jobs.

The positions we'll highlight in this chapter are **retail buyer** and **buyer/ merchandising trainee**, **sales service representative**, and **purchasing agent**.

What do these jobs have in common? First, they are all part of a network of jobs that provide information and services to support the movement of goods and services from the manufacturer to the ultimate consumer. This commercial area is an enormous part of our economy, and its workforce can potentially use as many math majors as might apply. In fact, if you were to talk to your college alumni office or career office about surveys taken of former graduates of your institution, you would probably discover a high proportion of math majors employed in these or the related occupations listed further on in this chapter. Let's look at each of these four representative jobs individually through the following categories.

Retail Careers

Retailing is the sale of selected merchandise directly to the consumer. Although this definition of retailing might be satisfactory for an exam, in actual practice the details of retailing fairly defy description! Retailing is an exciting career with enough change and involvement to keep you fascinated and on your toes for the rest of your career life. Retailing can also be a doorway through which you can enter the worlds of statistics, purchasing, product management, or training and development. Many top managers in these fields began their careers in retail buying.

Dramatic changes are taking place in the world of retailing. We will look at a few examples of this marketing revolution to give you a sense of the range, scope, and decision-making potential for the math major who is looking for a dynamic and fast-paced work opportunity.

High-tech marketing is not unusual in today's environment. Such marketing approaches are high cost and high risk but also have high potential for sales and profit. A career-minded retail professional must stay abreast of demographic trends and changing lifestyles and be ready to support high-risk decision-making efforts with relevant facts and analyses. Retailing often employs cutting-edge technology. Retail stores were the first to use videos on continuous loops in stores to stimulate customer awareness and interest in apparel products. Retailers were the first to sell in cyberspace and the first to use virtual reality to let shoppers "try on" clothes and experiment with the look of home furnishings or outdoor equipment in virtual environments.

Retailing is no place for the amateur or the faint of heart. It's a world that demands that owners take constant financial risk—and only the strong survive. One reason why the math graduate is a welcome candidate for jobs in the retail sector is that the continual, unrelenting competition of the marketplace demands superb abilities to analyze quantitative data and then make solid decisions based on that data. The first thing that greets most retail buying professionals each day is a comprehensive data printout indicating what has taken place in their store or chain the day before.

Do you wonder why you have not heard more about retailing careers before this? There are three principal reasons why math students don't have the opportunity to seriously consider retail careers upon graduation.

1. The subject of retailing is a science unto itself. It is serious big business that incorporates healthy doses of psychology, human behavior, and intuition along with high tolerance for taking risks.
2. Many students fail to consider retail, because traditionally the industry put no special premium on the attainment of a college degree. Of course, if you entered the executive suite of one of the country's large retailers, everyone you'd meet would have a college degree and so would most of the preeminent retail managers, sales managers, and buyers.
3. Students, parents, and faculty members may have a misperception of retailing in the hierarchy of American jobs.

Let's take a look at a typical college job fair. Math majors are going to be attracted to the large corporations, and the prestige and esteem of classic "management" positions, if not the more traditional statistics or actuarial positions. Retailers may also be represented at this job fair, but since math majors haven't had any in-depth exposure to retail careers in their major, all they're apt to see is store management and sales associate jobs. They don't see any potential there.

In reality, almost any of the retailers present at a job fair will offer careers with more latitude, more fiscal responsibility, more decision-making authority, and more opportunities to take off in a career than many other kinds of employers. Retailers may offer much more than many of the narrowly defined, computer-terminal-in-a-cubicle positions that exist beyond the impressive facades of corporate management positions. Retailing often provides talented people an opportunity to chart their own course, within the retailer's overall concept.

Merchandising is a term used to describe all the buying and selling activities within a store or chain. Merchandise managers decide what to buy based on what will sell. These are immensely complicated decisions. For example, if you are a buyer for men's shirts for a large chain, you must, perhaps a year in advance of the selling season, choose from among hundreds of fabric possibilities (cotton, nylon, silk, and blends), weaves (oxford, cambric, poplin, broadcloth, point-on-point, and twill), collar styles (button down, tab, many varieties of spread collars, and collarless), cuff styles (French or buttoned) and, of course, price points at which all these various shirt models will be sold. A budget for shirts alone in a large branch department store could represent hundreds of thousands of dollars a season. If you are thinking of spreadsheets, you're right on target!

Merchandisers who succeed and rise to the top in retail are people who have the best sense of what their market will buy—and who learn how to make the complex decisions that will satisfy their particular market. They develop this strong sensitivity in part by mixing with the market "on the floor." A good merchandiser stays in touch with his or her customers by working the floor; selling merchandise; hearing complaints and praise about products; and watching and studying the gender, age, and buying considerations of the public, both in person and through statistical and market research reports.

Retailing also embraces investment strategies, the costs of carrying inventory, pricing strategies, risk management, and many other planning tools that are highly sophisticated and require superb quantitative skills—in addition to decision-making ability and good communication skills, leadership, previous retail exposure, and the ability to juggle an ever-changing workload in a fast-paced environment. In retailing, what you can do and who you are count more than degree attainment.

Retailing is a field where you can jump right in and start a career if you are people oriented, service oriented, and willing to take advantage of opportunities. Retail personnel are on the front lines of getting the product to the consumer. They are the final marketing intermediaries. Customer contact and customer service are key to unlocking consumers' interest in purchasing your product. Customer contact provides the answers to ordering and pricing mysteries. Customer service is an art and a science, made up of both analytical and communication skills. It is these distinctions that keep your customers coming back.

Buyer/Merchandising Trainee

It is the job of a buyer to choose merchandise. Most buyers actually spend only a small percentage of their working year in the buying process. And, yes, for some shoes or leather articles, buying might take place in Italy; but more than likely, most buying will occur at a buying exposition for small leather goods in a convention hall in Atlanta or Dallas. Buyer jobs are hard work, and as you can see from the following ad, they are also highly quantitative in their orientation:

Assistant Buyer: Reporting to the buyer, the assistant buyer's role is to assist in achieving sales, gross margin, inventory turnover goals through proper merchandising management within company-defined guidelines. Focus in on the day-to-day operations of the department while learning long-range business strategy. Requirements: excellent PC, communication, organizational, and analytical skills are required—retail management training program a plus.

As a buyer trainee, you'll work under the supervision of a senior buyer. You'll need to learn to anticipate your customers' shopping needs several seasons before the merchandise is in the store, and you'll find yourself considering a variety of issues. Will men switch to a two-button suit? Will misses-size women tolerate a shorter skirt length? How many of each size children's shoe to order? The considerations go on and on. There are mountains of paperwork and sometimes daily reports of sales to analyze. And, if you haven't bought right, your mistakes stare back at you from the store shelves!

You'll learn to negotiate with vendors on costs, delivery dates, and shared advertising budgets. You'll work with store staff on merchandise displays, and as you gain experience, you will take on more responsibility in each of these areas. Before long, you'll be in charge while the senior buyer is away. Some organizations will require that your training include a stint in store management, so you may spend some time in a branch store.

Promotion to buyer positions can take two to five years. Realize, however, that the movement to middle management is faster than in many other industries. Economic ordering models are critical for most retailers and absolutely essential for the largest. The following position demonstrates an understanding that to get the best person for the job, they need someone with a math background:

Merchandising: Control Buyer. Seeking an individual to effectively manage the inventory of a category of goods so as to maintain an in-stock position on all key merchandise within planned inventory and sales levels. This highly visible individual will be a key liaison among Corporate Merchandising, Accounting, and vendors. Requirements: This position requires excellent verbal and written communications, analytical skills, and strong math/figures aptitude. PC knowledge required and forecasting/modeling experience a plus. Must be able to work well under pressure, be highly flexible, be detail-oriented, and have good organizational skills. College degree or related experience preferred.

Sales Service Representative

There's no question about it—sales is misunderstood as a career option and has certainly had its share of image problems. The reality is far, far different for the college graduate. As practicing career counselors, we hear two sides to the question "Is sales a career for me?" Students have one opinion of sales, which is generally not very positive, while employers have another view of sales, which is not only positive but exciting, challenging, and highly attractive. Well, since many of these employers were college students themselves not too long ago, you figure it out!

Teaching and counseling at a small public college has taught us that few college graduates view sales jobs as attractive. Many students have concerns about compensation, measures of performance, ethical issues surrounding product quality, and accuracy of sales information. We have the opportunity to meet and talk with professional salespeople throughout the year when they visit our campus to recruit, when we attend job fairs and professional conferences, and when we make site visits to employers.

Some of these sales professionals are new college graduates themselves, others are midcareer veterans, and yet others are senior staff who have long years of experience. They represent a full spectrum of experience in the field. Without exception, they speak of the professional challenges of larger accounts, important presentations to management, exciting travel opportunities, superb professional development opportunities, and satisfying financial rewards as well as the wonderful, interesting colleagues and business friends they have. Here's how they sum up sales:

- **Advantages:** A ready hiring market that allows you access to employment immediately upon graduation, the opportunity to acquire specific product information and become an "expert" in your field, and countless opportunities to learn and improve interpersonal skills that will remain valuable assets throughout your working life are just a few of the advantages to sales.
- **Disadvantages:** Not many. There is a strong emphasis on individual decision making, time-management skills, and self-direction. Increasingly, professional salespeople are required to be technically and quantitatively astute in order to manage a sophisticated use of technology needed in customer contact and service provision. They must meet the needs to be more of a listener than a talker, a problem solver than a solution dictator, and a consultant than a high-pressure order taker.
- **Return on investment:** Significant! A sales career will transform even the most broad-based business major into a real specialist in business affairs. You'll understand marketplace economics, consumer behavior, and organi-

zational systems in a way never possible in the classroom, and gain immeasurably in your ability to interact with individuals and groups in every type of setting. Sales is a jump-start to a career in which recognition comes faster than in most any other employment sector.

An Inside View

More interesting is how salespeople think of themselves. As a group, they are poised, professional, and comfortable in social situations. They would tell you, often as not, that a career in sales is the reason they are this way, not the other way around. Individually, they range from extroverted toastmasters to quieter, more scholarly types. Some are comfortable with large groups, others prefer one-on-one. Many think of their job as educational and informative, not persuasive and certainly not hard sell. They are deeply respectful of the people who comprise their market and do not see them as easily manipulated.

Selling Through Problem Solving

Contrary to public opinion, the objective of sales—certainly sales as practiced by professionals with a college degree and working for a reputable firm—is not to make somebody buy your product. A computer salesperson describes it this way:

> I see myself as a problem solver. I try to get inside and understand my customers' work, their needs, and their problems. I listen. After I've had an opportunity to think about it, I'll prescribe solutions to their problems. Those solutions will include my company's products, but sometimes, I will recommend a competitor's product if that is better suited to the job. I'm in this job for the long haul, and my belief is that, each time I sell only what the customers need, I build trust. The more trust I build, the less "selling" I have to do. If I continue to get good at what I do, I may never have to "sell" again.

Job Specifications

There are as many different types of sales positions as there are individuals to fill them. Each job holds the potential for both personal and professional growth, to varying degrees. Each job also places different demands on the job holder in terms of work productivity, self-management, travel, professional relationships, and product knowledge. Of special interest to a math major, some sales jobs will place a greater premium on his or her math education.

Let's identify some of the activities engaged in by salespeople. How they do their job is highly individual. With top sales professionals, you'll often see a singular professional style that is not only highly idiosyncratic but also indulged and approved of by top management. Glancing over this list, you can immediately understand that this job is far more complex and sophisticated than popular myth would have it.

- Identifying and contacting prospective customers
- Assessing needs and maintaining good relations with existing customers
- Designing and delivering sales presentations
- Keeping records/activity reports/sales performance records
- Tracking sales orders/delivery schedules and other details
- Handling complaints/returns when received
- Keeping an eye on the competition and reporting competitive activity
- Learning about new products and mastering marketing strategies

This short list of duties emphasizes communication, as you would expect, but there are many other skills and attributes that are suggested by this list. Let's identify and examine some of these other important skills and attributes close up.

Promotability

Any salesperson must understand the organization he or she works for in a very concrete, specific way. In the sales force, you learn how the organization is perceived by the consumer. This is true whether you are an admissions representative for a college or selling industrial boilers. Customer contact and your understanding and appreciation of the demands of a sales job will be the foundation and source of your credibility and authority as you advance in your career. One of the surest paths to promotion is having graduated from a corporate training program. Such training is usually exceptional, and it affords you an opportunity to meet many other people in the organization and gives you a broad overview of your employer.

Product Knowledge

You'll have to learn everything there is to know about whatever you're selling, whether it's corporate health plans, retirement systems, computer software, or large pieces of manufacturing equipment. Details, capabilities, costs, tolerances, competitive advantages, and a host of myriad details are crucial to being able to provide answers to questions. Product knowledge as a job demand can be confusing to those unfamiliar with retail. It doesn't mean an in-depth knowledge of what you prefer to buy but an awareness of what different publics want in consumer purchases. Those publics may be classified by income level, gender, or age.

Customer Knowledge and Contact

Sales is about meeting people—lots of people. Each of these persons has a different need and appreciation for your product. Most of the time you encounter your clients on the job. Since you have initiated the meeting, you will also encounter a variety of receptions, from a warm welcome to a glacial stare. It's going to be up to you and your sales skills to make these moments work. One way you'll learn to do this is by appreciating exactly what each customer wants. Is it service? Perhaps it's product quality or dependability. Sometimes it's the lowest price available. When customers are busy, you soon learn to judge each one as an individual, find out what he or she needs, and get down to business!

Communication Skills

No client believes you're interested in him or her if you're looking at your watch or tapping your foot while he or she speaks. Eye contact, your full attention, and the appropriate sounds and movements of affirmation and understanding convey the message "I'm listening to you!" Salespeople understand that communication is a complex business.

Sales is the art and science of communication. The communication is often about important issues such as product features, delivery dates, prices, conditions of sales, financing, and so on. A miscommunication can result in far more costly problems than just a lost sale. It's important that both parties understand each other. Professional salespersons become adept at ensuring that their message is correctly received and "decoded."

We've mentioned the importance of listening. Answering the client's concerns is also important. Ensuring that the client understands you can be accomplished through questioning, reframing and restating what you've said, and by writing things down. Who said salespeople were just great talkers? Salespeople need to be excellent writers, speakers, listeners, and nonverbal communicators!

Personal Attributes

People who are currently in sales or have enjoyed a sales job in their work history invariably speak of the "personal" skills they gained from sales work.

What are these personal skills, and why are they so important? We've identified some of the most critical ones.

- **Poise.** Meeting new people, ease in all social situations, and the ability to chat and make friends with a variety of people all develop an unconscious poise and confidence in sales professionals. It stays with them throughout their career, no matter where that is.
- **Ability to handle stress.** Sales situations—any situations involving people and negotiations—will involve stress as well. A sales career teaches the kinds of planning and strategies to anticipate and avoid stress, and the social skills to define and minimize the tension of stress-producing situations.
- **Time management.** Most sales positions demand exceptional time management. Learning which clients to call on during which times; deciding when to do your paperwork; determining how to best use driving and flying time; and strategizing your week, month, and year for best effect all develop excellent time-management skills. Many senior executives, when asked how they can be so productive, respond, "I began in sales and learned how to use my time effectively."
- **Decision making.** Whether you're on the road alone, in negotiations with a client, competing for a major account against a worthy adversary, or discovering new markets or sales opportunities, there will be a need to make decisions. Management knows you will frequently be called on to think for yourself and the good of the firm. Your job in sales will continue to demand good decision-making skills, many times on the spot.

Purchasing Agent

Purchasing agents usually work as part of a company's operations department. They buy things for companies, but unlike retail buyers, they don't buy items for resale to the final consumer. They buy all the raw materials, products, and services their companies need to maintain operations. They buy forklifts, industrial cleaning services, paper towels, spark plugs, drill bits, pencils, and industrial boilers. For manufacturers, they'll buy maple syrup, raw lumber, industrial diamonds, or granite, whatever the manufacturers need to produce their product.

Often working under the direction of a materials manager, the job of purchasing agents is far more complex than simple buying. They establish the sources of supply for their firms. In large organizations, this often means comparing bids, making in-house presentations to top management on planned purchases and schedules, visiting and monitoring vendors to ensure their operations are viable and that they can do what their representatives say they can do.

Because purchasing professionals buy so much product, they often can set prices and must be adept at the kind of calculations involved in adjusting quantities, discounts, and a host of other variables that affect pricing determination. In reading through this section, we hope you'll come to understand what an important job a purchasing agent has in an organization.

Purchasing agents are found not only in manufacturing but in all kinds of employment settings, for example, schools, hospitals, and government offices. Wherever they are located, their job is essentially the same. They ensure the company has sufficient supplies of the materials needed to continue operating. *Sufficient* is the key word here. If a purchasing agent orders too much, he or she has tied up operating capital that could have been used for other initiatives and may hurt overall profitability. Buying too much also subjects the stock to the risk of damage by deterioration, fire, theft, or simply being outdated by new and better products.

Purchasing agents are a crucial part of the management team and are recognized for the important contribution they make to the financial well-being of an organization. Not only do they have to know what to buy, but they have to make important decisions on the quality of the items they buy. Purchasing agents wouldn't be making a contribution by saving money on the purchase price of cheap items—poor-quality items wouldn't last very long and would have to be reordered frequently.

Purchasing agents make hundreds of decisions. For example, they must compute not only the cost of the item they are buying but how much it will cost to handle and transport the item. They must ensure that the vendors they are buying from can meet the promised delivery dates and quantity amounts, both of which can be critical. Knowledge of "just in time" and other current purchasing and contracting methods is needed to ensure that the organization gets the best delivery times and minimizes storage and shipping costs. You can save your firm money by having needed parts and materials arrive just when you need them and not before, but you can't do that without an enormous dependence on and trust in your vendors. Purchasing agents and vendors must have very close relationships because their livelihoods are mutually dependent.

Part of this relationship means the purchasing agent must coordinate his or her firm's production schedules with vendor production schedules and be completely familiar with transportion services and options. At one time, the United States experienced an unprecedented strike by a major transportation service's delivery drivers. This strike made a powerful impact not just on private citizens who had come to depend on that service but on many much larger organizations that discovered they had become too dependent on one delivery service. Following resolution of the strike, there were many articles in business and trade publications analyzing the impact of the strike and indicating that many purchasing agents had made significant changes in their use of delivery services for the future. Most of the changes involved distributing business among many delivery service vendors in order to weaken the effects of any future strikes on the cost of doing business.

It's the purchasing agent's job to anticipate problems such as that posed by transportation strikes, extreme weather conditions, and other things that can impact delivery times, and solve the problems that come with them. To do this, purchasing agents work closely with many other departments in their organization, including receiving, traffic, and supply departments.

In small firms, one purchasing manager may do everything. In larger firms, such as a government weapons contractor, for example, there may be more than a hundred purchasing agents, each specializing in a certain class of material or machine.

Some purchasing agents begin as expediters and work their way up. Expediters handle much of the paperwork and other details involved in making purchases, arranging shipping, and settling claims. Their responsibility is to see that delivery commitments made by the vendor are kept or, if delays do occur, figure out ways to speed things up. As an entry-level job in purchasing, it involves a lot of paperwork, although the job has expanded as firms seek to reduce the time between orders and delivery. It's a great way to learn the business.

As an expediter, you'll have an opportunity to become familiar with most items purchased as well as who purchases them—that's the kind of information you'll need to move up in this field. Some entry-level positions are simply designated "trainees" within a purchasing department. Trainees learn on the job and may attend special classes. Often, this trainee period will include a rotation through other departments of the organization as well.

Working Conditions: Retail Buyer and Buyer/Merchandising Trainee

Buyers and trainees usually work in pleasant office environments, within a team of other employees in marketing, sales, and other related departments. Hours are usually regular but occasionally will require overnight, weekend, and other travel, plus occasional long hours in the home office on a seasonal basis.

Talent is immediately recognizable in the fast-paced and highly responsive world of retailing. Ambitious people usually want positions that demonstrate their talent to advantage and then reward that talent with responsibilities and duties that will further stretch and develop their skill.

Some people love to be busy, the busier the better. They cannot tolerate sitting at a desk all day, chatting around the water fountain, or catching up on business journals. The answer is retailing. Each day has a different outcome, as each customer through the door presents a different collection of needs to be satisfied. Demands can be incessant. This is a business for the strong and energetic. If you're not happy in such an environment, you'll see it as chaotic, frantic, and frustrating!

Sometimes job applicants make the mistake of telling the interviewer, "I'm a people person." What does that really mean? In retailing, a "people person" must be comfortable with all kinds of people from all walks of life and, increasingly, from many different parts of the world. You will have to want to understand their needs and enjoy making an effort to fulfill those needs.

Retailing careers offer unparalleled mobility. Retail opportunities exist worldwide and in every size and variety of emporium. So, if you are anxious to live on the West Coast or in the Plains States, you can do it.

The word *entrepreneur* has special meaning for retailers because, in a sense, every buyer is in business for him- or herself. Each department or operating unit in today's retail stores has its own printout of profit and loss. You will easily be able to demonstrate your entrepreneurial spirit.

Take a look around. Retailers are open holidays (even some of the most sacrosanct, such as Christmas), evenings, weekends, and late at night. Stores need to be staffed, maintained, and managed. You'll be working many times when the rest of the world is out playing! Competition is intense in retail, and management often puts in long hours on and off the floor to stay abreast of the competition.

Even when you obtain the coveted position of buyer, you'll be shocked to find that your office may be just a cubbyhole tucked behind the dressing rooms on the fifth floor. Store floor space is used for merchandise. Employees usually have to take what's not usable or what's left over, and any glamour is reserved for the customers.

Working Conditions: Sales Service Representative

Sales jobs are plentiful, but as this path indicates, they vary widely in quality, professionalism, training provided, use of your math background, opportunities to interact with colleagues, compensation, and growth potential. When you are a new grad in the market for a sales job, there's a temptation to look at all the sales jobs available, rationalize away any concerns you might have, and take the first one offered you.

Three aspects of most sales jobs are so dominant that they serve to define the nature of this employment experience, and you would be well advised to consider your suitability for a sales position against these criteria: degree of sociability, nature of the client contact, and productivity/competitiveness. Go back to Chapter 1 on self-assessment, and review what you learned about your personal traits, values, and skills, and then measure what you've learned about yourself against these three critical aspects of sales positions.

Most people remark on how easily sales professionals talk to just about anybody—the postal carrier, the cashier at a supermarket, the man or woman on the street. It's quite easy to explain. Despite the fact that their job emphasizes connecting with people, most sales professionals work alone, and their contact with others is fairly brief and intermittent. They spend quite a bit of time alone, and any one of them would tell you that you need to be comfortable with yourself to be good at sales.

In most sales jobs, the actual "selling" of a product is not as challenging as connecting with the client. There are basically two ways to contact a client: (1) through some sort of prescreening or presales work that has the client requesting your visit, or (2) by "cold calling," where you call on a client who may not yet be aware of your product or service but who you believe might be a good prospect for it.

Whether being invited or inviting yourself is more appealing or more challenging to you has much to do with your personality. Your tolerance of risk, ability to handle new situations, and ease in making acquaintances are good barometers of which situation would be best for you over the course of a career.

Sales is often repetitive. It's not a career where you can sit home and rest on your laurels. For many in this profession, success only motivates them to further achievement. This competitive drive may not be directed at others. It may simply be a competition with yourself. Because sales always involves seeking new markets and new clients for products and services, that competitive spirit is essential.

Working Conditions: Purchasing Agent

Although most of your time is spent in an office, purchasing department employees will also often visit plants and warehouses, and attend meetings, exhibits, and conferences. Whether or not your workweek is fairly standard is largely dependent on your particular sector of employment. There are many industries that, because of intense seasonal production, demand overtime beyond the standard workweek during periods of intense output. Purchasing managers find they are called on to solve problems, handle details, and coordinate many activities during such intense work periods.

Training and Qualifications: Retail Buyer and Buyer/Merchandising Trainee

The retailing environment is less one of glamour, travel, and clothes than one of computer printouts and statements of profit and loss. Because computerized information can tell the retailer so much about what is required in terms of inventory, labor, and so on, students seriously considering retail careers need to consider mastering computer technology, including spreadsheet software and statistical packages, and have some exposure to database management.

As our economy becomes increasingly technologically oriented, the candidate who can apply high-tech, information-based skills to a successful career in retailing has many advantages. In addition to these technical skills, there is a need for analytical skills. A course in marketing, general management, and/or economics might give you an even broader picture of the changing and very dynamic scope of retailing in our economy. The appearance of increasingly larger merchandisers, whether they be department stores, warehouse showrooms, or grocery retailers, places demands on those candidates to understand the mechanics of big business.

Internships or, if your school has one, a co-op program is an ideal way to help you ascertain your interest and suitability for the retail profession. Additionally, they help you to add valuable entries on your résumé that will attract employers, following graduation. Part-time employment in any area of the retail sector will give you immense credibility as a manager who can truthfully say to rank-and-file workers, "I've done that."

Training and Qualifications: Sales Service Representative

Many are called, but few last. The potential for success in a sales position often can't be determined well during the interview process. It may take several months of on-the-job work for a person to assess whether he or she is right for sales. Some of the best sales organizations hope to improve their odds by having you interview with a large number of people to get a variety of impressions about you and your potential.

Following selection, the very best sales firms have you undergo a lengthy (up to six weeks) training period, which will involve role-playing, testing, and acquisition of product knowledge. These training programs are usually very professional, and you'll learn lessons that will stay with you throughout your entire working career. Some firms train you full-time, others incorporate the training into part of the workday, and yet others have you begin work and then have you return to training. The philosophy behind this last option is that the worker who has already experienced the realities of the job will understand the value of the training more readily.

Training and Qualifications: Purchasing Agent

Entry into the field of purchasing is highly competitive, and a college degree is a must. In general, entry-level candidates with a degree in quantitative areas such as mathematics and computer science, and business degrees that focus on operations management are highly desirable.

A number of certification programs are available for purchasing professionals. In private industry, the hallmark of experience and professional competence is the Certified Purchasing Manager (CPM), which is awarded by the National Association of Purchasing Management, based on criteria involving years of experience and successful completion of a series of examinations. Certified Purchasing Professional (CPP) and Certified Purchasing Executive (CPE) designations are available through the professional development efforts of the American Purchasing Society. In governmental and public employment, Certified Public Purchasing Officer (CPPO) is a designation awarded by the National Institute of Governmental Purchasing, after the candidate meets similar testing and experience requirements. It goes without saying that achieving this notable designation is enhancing to your career and your bank account.

These designations of excellence are a sign of how professionally the purchasing field handles itself. Continuing education is the norm, and, if you are located in a high-tech field, then continuing professional development will be imperative. Be certain to ask during your interview process what provisions are made for your training and continued professional growth by your employer. You'll be glad you did.

Certain personal qualities and characteristics are essential to success in this area: high ethical standards; good analytical skills; and strong communication skills, coupled with good listening skills.

High ethical standards are critical for people who hold these influential positions in a firm. Imagine the responsibility that comes with such buying power and the ability to award contracts that represent huge sums of money. Purchasing agents are under incredible pressure from agents, salespeople, representatives, and vendor executives. There may be offers of kickbacks, special favors, and, more frequently, just simple requests for you to overlook something.

If you decide to enter this field, you must make your ethical standards clear and out in the open to everyone, including your colleagues and representatives of vendors. The field is full of stories of talented people who've lost their jobs because they thought they could bend the rules. Don't be one of them.

Analytical skills are absolutely essential. Because purchasing professionals spend much time analyzing technical data in suppliers' proposals and making complicated buying decisions that cost large sums of money, they must be comfortable with complex data and have the ability to see that data from a number of perspectives. In many cases, analysis comes down to the ability to creatively employ a variety of perspectives on a problem in anticipation that one of these viewpoints will suggest a solution.

Ethics and analytical skills are not the only important personal qualifications—good communication matters. Purchasing professionals are valued for their ability to communicate clearly and effectively. This is an area where no means no and needs must be clearly articulated. Purchasing agents work closely with their vendors and develop strong business relationships. Consequently, language is critical, and there can be no suggestion of promises when promises are not intended. So your ability to communicate well—and precisely—will be an important hiring criterion. You may want to polish your interview skills and refer to the opening chapters of this book.

Personal strength goes hand in hand with excellent communication and high ethics. Employers look for job candidates who know who they are, what they want, and how they function effectively. A positive self-image and a preference for variety, challenge, and professional growth will be some of the characteristics sought after in any interview for purchasing jobs.

Earnings: Retail Buyer and Buyer/ Merchandising Trainee

Wholesale and retail buyers, other than those in farm products, earned a median income of $40,780 in 2002, according to the *Occupational Outlook Handbook*. The middle 50 percent earned between $30,040 and $55,670 a year, while the lowest 10 percent earned less than $23,270 and the highest 10 percent earned more than $76,070. Median annual income for purchasing agents and buyers in farm products was comparable, at $40,900.

If you are in a trainee role, your salary is apt to be somewhat less than a designated assistant buyer's or assistant merchandise manager's. Depending on the employer's size (in terms of sales volume) and number of employees, you can expect a salary range of $21,000 to $24,000. Promotions out of the trainee role will bring a marked increase in your pay. It's also important to note that benefits in these positions tend to be very competitive, and a good benefits package can add up to a third of your salary in value.

Medical insurance benefits are in a state of flux currently, and many companies are asking employees to pay more of the cost; so you need to check carefully to see what is being offered in these packages. You may not realize the importance of your benefits package until you visit an emergency room or have major dental work done, but if that should happen, you'll begin to discover the "hidden income" of your benefits package.

Earnings: Sales Service Representative

Sales service representatives' earnings frequently can offer the entry-level, new-graduate applicant the highest possible earnings available with a college degree, with a few exceptions for some technical majors. High earnings are attractive to many students who graduate with some debt burden and a strong motivation to begin to live on their own and make a life for themselves.

Positions in sales will pay either a salary, a commission, or a combination of the two. You must consider compensation plans carefully regardless of your financial needs. It can be very disheartening to leave a position because of a disagreement over pay arrangements when a little forethought and detailed discussion might have anticipated problems.

As an entry-level salesperson, you will probably be more attracted to a salaried selling position. It is less risky and puts less responsibility on you to produce than a commission system does. Straight salary plans work best for employers when it is difficult for management to determine which person on the staff actually made the sale or when the product involved has a broad, cyclical sales pattern, which would leave the sales staff with virtually no income during the slow periods if only commissions were involved.

Sales professionals who have gained skills and confidence in their abilities are often more attracted to commission plans. Commission plans offer them unlimited income if they are successful. Straight commission gives the greatest incentive to salespeople, while maintaining a predictable sales cost in relation to sales volume. Two factors working against straight commissions are (1) high turnover and burnout of sales staff, and (2) a sales process tainted by pressure and the need for sales personnel to "move" product to pad the salesperson's paycheck regardless of his or her customers' true need for the products or services.

Combination plans exist with fixed salaries and incentive features added on. This builds continuity in the sales force, yet allows superior sales staff to shine, and encourages everyone to develop more business.

Each pay plan has its own advantages and disadvantages. A straight salary job cannot reward superior achievement, which may become frustrating if you are exceptionally successful in your position. If you expect to be the best you can be, shouldn't your paycheck reflect that? With a straight salary job, you and the poorest performer in the team may be taking home the same paycheck, at least for a time.

Commission rewards ability and performance, which can be attractive to the confident sales pro. However, be sure you understand what the minimum performance levels are, how much you need to do to earn your commission, and how feasible those goals may be. Pay periods can vary also. What is the waiting period, following a successful sale, before you will receive your compensation?

Expenses can be another area of difficulty. Some companies expect their employees to pay their own business expenses out of pocket and then wait to be reimbursed. If the company doesn't reimburse you in a timely manner, you could experience a cash shortfall when bills come due. Be sure you know ahead of time what the company's policy is regarding reimbursement of expenses for out-of-town company trips.

Commission is not as risky as most recent graduates may think. If (and this if is important) the organization that hires you is spending money and time on your training, its managers do not want to see you fail and are taking steps to ensure you succeed. These compensation plans are based on experience born of what past salespeople have done and what incentives they needed to do it. There is no success if salespeople drop out because they can't make any money. The organization would suffer and, ultimately, fail. Be sure you have financial backup, but don't be unnecessarily wary of working on commission.

Earnings: Purchasing Agent

Purchasing agents are well paid, and median annual earnings in 2002 were $45,090. The middle 50 percent earned between $34,820 and $58,780. The lowest 10 percent earned less than $27,950, and the highest 10 percent earned more than $73,990.

Purchasing managers earned a median annual income reported at $59,890 in 2002. The middle 50 percent earned between $43,670 and $81,950 per year, with the lowest 10 percent earning less than $32,330 and the highest 10 percent earning more than $108,140.

Because of the nature of the firms they work in (size, number of employees, and income production), these positions would, in most cases, receive excellent benefits packages, including medical, dental, life, and health insurance; vacation time; and sick leave, although many companies are paying less of the total cost of medical insurance or are otherwise limiting the kinds of medical benefits provided. Bonuses are a common part of compensation packages for these jobs and can represent a significant boost to income.

Career Outlook: Retail Buyer and Buyer/ Merchandising Trainee

Overall, new job growth in this area is expected to be slower than average through 2012, because of the overall state of the economy, according to the *Occupational Outlook Handbook*, although some sectors may still show growth.

Career Outlook: Sales Service Representative

The outlook for sales service representatives is somewhat better. This group is expected to grow about as fast as the average through 2012. Note, however, that much depends on the particular industry and its trends.

As with all the jobs in this path, turnover is high, and as people leave sales for other occupations, new openings occur. You'll find this turnover highest among those firms that offer the least training and whose products are the least sophisticated. Not much has been invested in the employee, nor has the employee invested much in the employer. As a result, leaving is easier.

Career Outlook: Purchasing Agent

The projected slower-than-average growth for jobs in purchasing as predicted by the U.S. government is largely a result of the computer. Consolidations, mergers, and acquistions have also played an important role. As purchasers write longer-term contracts with larger organizations that can supply more products and services from one location, it dramatically reduces the time normally spent on negotiations and the personnel traditionally involved with that.

On the public purchasing front, new laws affect purchasing employees. In 1994, the federal government mandated electronic purchasing for all items under a specified dollar amount mandated by the Federal Acquisition Streamlining Act (FAST), and that regulation continues to apply today. FAST and recent federal budget cuts have greatly reduced the demand for new hires in federal purchasing jobs.

Minimum-paperwork buying occurs in more than just the government sector. The purchasing agent's role has changed dramatically because of the computerized use of credit cards within companies and other organizations. Until recently, everything purchased was processed through the purchasing office. Now, with the issuance of corporate purchasing credit cards, individual employees are often able to make direct corporate purchases in defined areas such as at conferences and conventions, and over the Internet. Innovations such as this kind of "direct buying" have reduced the number of purchasing agents needed in the marketplace.

Strategies for Finding Jobs: Retail Buyer and Buyer/Merchandising Trainee

Retailing is an incredibly diverse field. Merchandise lines run the gamut from lingerie to lawnmowers, with big names in each field. Geographically, retailing is literally all over the map, and you can find retail employment throughout the United States and the world. So, in addition to many choices of product lines (merchandise) and location options, there are equally as many choices in job duties: buyer, merchandise manager, store or department manager, site development, advertising, finance—and many more.

The following information will be useful in helping you develop a successful strategy for finding the right job in the retail sector. If you have enough time to do a thorough job search, have access to all the options that follow, and want to be sure you've covered all possibilities, then go through this plan, step by step. Not only will you succeed in finding the jobs, but the process of implementing this search will also make you a better and more informed interview candidate, because you will have increased your understanding of the world of retailing.

Campus clubs that bring students together for common interests are wonderful sources of information about a possible career in retailing. Many schools offer a campus branch of the American Marketing Association, which will welcome math majors who have an interest in their field. Try to meet every guest speaker who comes to campus to speak to the club on business issues. Let them know of your career plans as a math major interested in retail, and ask their advice. Serve on club speaker committees and any special career-related projects. If your club has an opportunity to attend a national convention, try to be part of your school team and use that opportunity to meet guests, gather information, and network.

Your college may provide an opportunity for you to elect a work-study plan called cooperative education. If you can manage it (it may add anywhere from a semester to a full year to your degree), cooperative education is a proven way to access a career, learn the ropes, meet influential people in your degree path, and be exposed to the dimensions of your chosen field while you earn some money for tuition.

The single most effective way to improve your ability to move from graduation to a job is through an internship. Internships are generally nonpaid or low-paid training positions in your chosen field that give you broad exposure to a variety of tasks and management role models. They can last from a few weeks to a semester and sometimes longer. Internships can often be used for college credit toward your degree. Many successful interns have been offered positions at the organization sponsoring the internship upon graduation. Many others credit their job success to their internships.

The alumni office on your campus can put you in touch with former graduates of your college who are now working in the retail job market. These alumni connections can be very helpful, offering informational interviews, background on the firms that employ them, and general insights into the retail job market. Depending on the sophistication of your school's alumni database, it my be possible to actually isolate who's working as a buyer, a merchandiser, or store manager, including other former math majors.

Remember that when you contact these individuals for career guidance, you are representing your college and every other student who may someday want to use this valuable referral service. Be prepared with a list of questions for your alumni contact, and use his or her time wisely. You'll make a good friend and have increased insight into your chosen career field. Many alumni will invite you (distances permitting) for a visit to their places of business and may offer to assist with your résumé or job search strategy.

College intranet computer listings (or actual three-ring binders) that contain some of the hundreds of job advertisements that come through the mail to your career office on campus can inform you of many excellent retail entry-level jobs. These may be duplications of the advertising copy that was placed into a newspaper classified want ad section, or they may be a special recruitment mailing sent to colleges for qualified applicants. These job postings are easy to scroll through, and, because they change frequently (new ones arrive weekly), you should make this kind of "catalog shopping" a weekly activity. Many colleges provide this service electronically from off-site locations, utilizing a password or student identification number.

Many of the professional associations listed at the ends of Chapters 6, 7, 8, and this chapter provide job listings via their websites. In addition to the campus career office website with its job bank, there may be résumé databases for students that offer links to websites such as Jobtrak (jobtrak.com), a job database targeted to college students and alumni. These job banks allow you to search through databases of openings; résumé databases where you can post your résumé online; career information services that offer advice on résumé writing and job search techniques; and virtual career centers, which offer you two or three of these services. Check with your career center. They may recommend some tried-and-true websites that can speed up your search. Other listings may be found through websites such as the following:

- *The Riley Guide*, developed by Internet job search and recruiting consultant Margaret Riley Dikel (rileyguide.com)
- *JobWeb*, provided by the National Association of Colleges and Employers, especially for college students and new college graduates (jobweb.com)
- *USA Jobs*, operated by the Office of Personnel Management, the federal government's official jobs site (usajobs.opm.gov)

In addition, *Hook Up, Get Hired!: The Internet Job Search Revolution* and *Electronic Job Search Revolution* by Joyce Lain Kennedy are two publications that give the reader ways to search for jobs and reach potential employers through the Internet.

Sites on the Internet can be searched by occupational title, such as buyer, merchandise manager, purchasing agent, or sales representative. These sites can be searched geographically as well. The Government Job Finder lists a publication, *Purchasing Notes*, published by the National Institute of Governmental Purchasing, that posts job vacancy listings to its members at nigp.org.

Employers from the retail sector who visit your campus are giving you a strong indication of their interest in your school's students. Sign up for every recruitment interview with a retailer that you possibly can. On-campus interviewing can and does lead to actual job offers. It's excellent interview practice as well. More important, you will begin to develop a sense of what each employer is offering and start to make distinctions about what you feel is the best "fit" for you in a retail career offering. Unlike the job fairs, these are private interviews, one-on-one with a senior representative from the retail organization. Even better, it's held on campus in familiar surroundings, which should prove helpful to you in controlling those interview jitters. Your campus career office probably maintains files on all the recruiting firms to allow you to be thoroughly prepared for your interview.

Job fairs are valuable job search tools for the student seeking a retail career opening. First, they are very efficient. How else would you be able to meet and talk with so many possible job contacts in one day? Second, retailers use them! Rosters of employers at past job fairs indicate that retailers are traditionally very well represented. Retailers enjoy job fairs because it allows them to meet and see a number of highly qualified majors. Third, the job fair process of walking up to an employment representative, greeting him or her, and giving a one-minute "infomercial" about who you are and what you might have to offer his or her organization is a perfect example of the specific kind of people skills retailers seek out.

If a recruiter at a job fair is interested in you, he or she may ask you to submit a formal application, or you may be invited to one of the employer's "open houses" or an actual interview. The recruiter may extend this invitation at the job fair or by telephone or letter at a later date. Strong student job candidates who take full advantage of job fair opportunities are consistently amazed at both the number of responses and how much later after the fair some of them arrive.

Strategies for Finding Jobs: Sales Service Representative

Your job search approach may be your ticket to a job! There's something interesting about the hiring of sales personnel that's different from the hiring of other kinds of professionals. With retail buyers, merchandisers, sales service representatives, analysts, purchasing agents, or administrative support, the employer looks for professionalism in the résumé and cover letter, but that is not the focus. They will be more interested in finding out what the job candidate knows, what his or her experience has been, and whether he or she seems adequately prepared for the job under consideration.

In the sales personnel market, every aspect of the coming together of the candidate and potential employer provides a clue to the candidate's potential in a sales job. What better exercise in persuasiveness is there than the job search? Can you think of any other activity that puts as much emphasis on the kind and quality of one's communication as the job search and the interview itself? A job search is by and large a sales exercise, and if you wish to become a sales professional, an employer has every right to expect to see the beginnings of brilliance in how you go about seeking your job!

Strategies for Finding Jobs: Purchasing Agent

Much of the general job search information outlined for both the retail and merchandising positions and sales service representative holds true for the purchasing agent job search. There are some qualitative differences in the search, however. While a math degree is valued but not essential for jobs in retail and sales, it is a far more important résumé item for a purchasing job. Additionally, since at any one time there are fewer purchasing jobs than retail or sales positions (which number among the most populated jobs), your candidacy is going to be closely scrutinized, because it will be more competitive.

As a math major, you will probably have had little, if any exposure, to purchasing careers during college. That's okay. Most purchasing professionals learned their trade on the job. And while purchasing agents' jobs can be remarkably similar, with allowances for both the size of the firm and nature of the work they do, you may want to begin your search by focusing on two or three of these possible employment sectors. For example, you may decide that you want to work in a setting such as a hospital, educational facility, or corporate hotel firm where you would be buying finished goods. Concentrating on these three sectors will help you focus your interview skills, add specific vocabulary and concerns that recur in your interviews, and generally allow you to better target your search. You could just as easily focus on the aircraft industry, defense contractors, or the automobile subcontractor sector. Most firms will prefer a bachelor's degree candidate with an emphasis in quantitative or business skills.

Possible Job Titles

The variety of titles for the jobs in this field are nearly infinite, as you will see as you progress through the ads in your job search. In a short time, you will get a feel for the terms used and be able to recognize even some obscure combinations of key words. Listed below are just a few of the possible titles that you will encounter for a retail buyer and buyer/merchandising trainee or a sales service representative.

Assistant buyer
Buyer
Buyer/merchandising trainee
Buyer trainee
Department manager
Merchandise manager
Retail buyer
Sales associate
Sales service representative
Store manager

The sales industry works very hard to improve the professional standing of its salespeople and remove any stigmatizing labels or job titles that might inhibit their ability to do their job. So you'll see several titles that don't even contain the word *sales*. Get used to that in seeking out sales positions. It may be your first signal that the hiring organization has paid particular attention to the role of its sales force. Following are some examples of titles:

Account executive
Account representative
Area director
Area manager
District manager
Major account representative
Market representative
National sales manager
Outside sales representative
Product line manager
Product manager
Purchasing agent

Regional manager
Sales consultant
Sales counselor
Sales director
Sales manager
Salesperson
Sales representative
Sales specialist
Service representative

Possible job titles for those beginning in the field as a purchasing agent include the following:

Assistant buyer
Expediter
Junior buyer
Purchasing clerk
Trainee

Possible job titles for a purchasing agent in government include the following:

Agent
Buyer
Commodity purchasing manager
Contract specialist
Contract supply manager
Director of contracts, pricing, and procurement
Director of corporate materials
Director of materials management
Inventory manager
Manager of materials acquisitions
Manager of purchasing
Materials analyst
Procurement team leader
Purchasing director
Purchasing engineer
Strategic planner
Traffic manager

Related Occupations

Related occupations include the following:

Buyer/merchandising trainee
Comparison shopper
Insurance agent
Manufacturer's representative
Procurement services manager
Retail buyer
Retail sales worker
Sales service representative
Services sales representative
Traffic manager
Wholesale sales representative

The skills you've acquired in sales tend to be universally esteemed: listening, speaking, public presentation, decision making, analyzing, time management, product knowledge, and an ability to "connect" with people authentically and quickly. Sales professionals are easily transitioned to a variety of occupations. A sampling includes the following:

Advertising
Career counseling
Communications
Consulting
Insurance sales agent
Manufacturer's representative
Marketing and advertising
    manager
Marketing management
Materials manager
Procurement services manager
Product management
Promotions
Public relations
Purchasing agent
Real estate
Retail sales worker
Sales manager

Services sales representative
Trade fair design
Traffic manager
Training and development
Travel agent
Wholesale sales representative

Professional Associations

Following are some of the associations that relate to retail buyers, buyer/merchandising trainees, sales service representatives, and purchasing agents. For more information about these professional associations, either check the website listed or use the *Encyclopedia of Associations* published by Gale Research, Inc. Review the "Members/Purpose" section for each organization and decide whether the organization is related to your interests.

Membership in one or more of these associations would be helpful in terms of job listings, networking opportunities, and employment search services, as well as your ongoing professional development. Some associations provide information at no charge. If you want to receive their publications that list job opportunities, you may be required to join the association. Check for student member rates. Check the websites for further career information or links to other sites that provide information related to this chapter's occupational paths. Investigate the associations to see how they might help you in your job search.

American Marketing Association
311 S. Wacker Dr., Suite 5800
Chicago, IL 60606
ama.org
Members/Purpose: Professional society of marketing and marketing research executives, sales and promotion managers, advertising specialists, and others interested in marketing.
Journals/Publications: American Marketing Association proceedings, international membership directory, books, courses, online seminars

American Purchasing Society
430 W. Downer Pl.
Aurora, IL 60506
american-purchasing.com

Members/Purpose: Seeks to certify qualified purchasing personnel. Maintains speakers bureau, library, and placement service. Conducts research programs; compiles statistics including salary surveys. Provides consulting service for purchasing, material management, and marketing.

Journals/Publications: *Cost Cutter*, directory of buyers and purchasing executives, journal, *Professional Purchasing*, articles, books, online career center

Association of Retail Marketing Services
10 Dr. James Parker Blvd., Suite 103
Red Bank, NJ 07701-1500
goarms.com

Members/Purpose: Devoted to the promotional needs of the retail industry. Recommends incentive promotion at the retail level.

Journals/Publications: ARMS membership directory, directory of top fifty wholesale grocers

Federal Acquisition Institute
U.S. General Services Administration
c/o Defense Acquisition University
9820 Belvoir Rd.
Fort Belvoir, VA 22060
fai.gov

Members/Purpose: FAI is part of the U.S. General Services Administration (GSA), which acts as the executive agent for the FAI. The Office of Federal Procurement Policy, Office of Management and Budget, is responsible for providing for and directing the activities of the FAI. The FAI provides expertly managed space, supplies, services, and solutions to enable federal employees to accomplish their missions. Work space, security, furniture, equipment, supplies, tools, computers, and telephones are also provided. The FAI is found on the GSA's website and offers online training, educator's conference, online resources, federal acquisition regulations, bulletin boards, forums, publications, associations, legislative links, and judicial links.

Journals/Publications: See website.

Hospitality Sales and Marketing Association International
8201 Greensboro Dr., Suite 300
McLean, VA 22102
hsmai.org

Members/Purpose: An international organization devoted entirely to education of executives employed by hotels, resorts, and motor inns.

Journals/Publications: Directory, *Marketing Review*

Institute of Store Planners
25 N. Broadway
Tarrytown, NY 10591
ispo.org
Members/Purpose: Persons active in store planning and design, visual merchandisers, students and educators, contractors and suppliers to the industry. Dedicated to the professional growth of members while providing service to the public through improvement of the retail environment.

Journals/Publications: *Directory of Store Planners and Consultants, ISP International News*, newsletter

Institute of Supply Management
P.O. Box 22160
Tempe, AZ 85285-2160
ism.ws.org
Members/Purpose: Purchasing and materials managers for industrial, commercial, and utility firms; educational institutions; and government agencies. Issues reports on market conditions and trends; disseminates information on procurement; works to develop more efficient purchasing methods; conducts programs for certification as a purchasing manager; operates placement service.

Journals/Publications: *International Journal of Purchasing and Materials Management, NAPM Insights, Report on Business*

Museum Store Association
4100 E. Mississippi Ave.,
Suite 800
Denver, CO 80246-3055
museumdistrict.com
Members/Purpose: Sales departments in museums, including museums of fine arts, history, ethnology, and science. Encourages dialogue and assistance among members.

Journals/Publications: Membership list, *Museum Store, Product News*

National Retail Federation
325 Seventh St. NW, Suite 1100
Washington, DC 20004
nrf.com
Members/Purpose: Retailers of men's and boys' clothing and furnishings.
  Maintains government liaison.
Journals/Publications: *Better Retail Salesmanship*, members newsletter

National Association of Independent
  Publishers Representatives
111 E. Fourteenth St.
New York, NY 10003
naipr.org
Members/Purpose: Independent publisher's representatives selling
  advertising space for more than one publisher of consumer, industrial,
  and trade publications.
Journals/Publications: Bulletin, roster of members, employment
  opportunities online, sample contracts

National Association of Men's
  Sportswear Buyers
309 Fifth Ave., Suite 307
New York, NY 10016
namsb-show.com
Members/Purpose: Sponsors trade shows for buyers of clothes for
  menswear stores. Conducts media interviews to discuss menswear and
  educational programs.
Journals/Publications: Newsletter

National Association of State
  Procurement Officials
167 W. Main St., Suite 600
Lexington, KY 40507
naspo.org
Members/Purpose: Purchasing officials of the states and territories.
  Council of State Governments serves as staff agency. Operates
  "Marketing to State Governments" seminar.
Journals/Publications: *NASPO News, Contract Cookbook, State and Local
  Government Purchasing, How to Do Business with the States: A Guide for
  Vendors*

National Contract Management Association
8260 Greensboro Dr., Suite 200
McLean, VA 22102
ncmahq.org
Members/Purpose: Individuals concerned with administration, procurement, acquisition, negotiation, and management of government contracts and subcontracts. Works for the education, improvement, and professional development of members and nonmembers through national and chapter programs, symposia, and workshops; develops training materials to serve the procurement field. Offers certification in contract management (CPCM and CACM) designations.
Journals/Publications: *Contract Management, National Contract Management Journal*

National Institute of Governmental Purchasing
151 Spring St.
Herndon, VA 20170-5223
nigp.org
Members/Purpose: Federal, state, provincial, county, and local government buying agencies; hospital, school, prison, and public utility purchasing agencies in the United States and Canada; develops simplified standards and specifications for governmental buying; promotes uniform purchasing laws and procedures; conducts specialized education and research programs; conducts certification program for Professional Public Buyer and Certified Public Purchasing Officer; offers consulting services and cost-saving programs and tools for governmental agencies, including purchasing software for desktop computers; maintains specifications library for public purchasing.
Journals/Publications: *National Institute of Governmental Purchasing— Letter Service Bulletin, NIGP Dictionary of Purchasing Terms*, technical bulletin

National Minority Supplier Development Council, Inc.
1040 Avenue of the Americas, Second Floor
New York, NY 10018
nmsdcus.org
Members/Purpose: Minority businesspersons, corporations, government agencies, and other organizations who are members of regional purchasing councils or who have agreed to participate in the program. Regional councils certify and match minority-owned businesses with

member corporations that want to purchase goods and services. Conducts sales training programs for minority entrepreneurs and buyer training program for corporate minority purchasing programs.

Journals/Publications: *Minority Supplier News*, minority vendor directory, National Minority Supplier Development Council annual report

Promotion Marketing Association
257 Park Ave. South, Suite 1102
New York, NY 10010
pmalink.org

Members/Purpose: Promotion service companies, sales incentive organizations, and companies using promotion programs. Supplier members are manufacturers of premium merchandise, consultants, and advertising agencies.

Journals/Publications: *Outlook*, membership directory, Promotion Marketing abstract

Radio Advertising Bureau
1320 Greenway Dr., #500
Irving, TX 75038
rab.com

Members/Purpose: Membership includes radio stations, radio networks, station sales representatives, and allied industry services such as producers, research firms, schools, and consultants. Exhorts advertisers and agencies to promote the sale of radio time as an advertising medium.

Journals/Publications: *RAB Instant Background: Profiles of 50 Businesses*

Sales and Marketing Executives International
P.O. Box 1390
Sumas, WA 98295-1390
smei.org

Members/Purpose: Seeks to make overseas markets more accessible by interchange of selling information and marketing techniques with executives in other countries.

Journals/Publications: *Marketing Times*, leadership directory

Index

What can I do with a degree in math?

You've worked hard for that math degree. Now what? Sometimes, the choice of careers can seem endless. The most difficult part of a job search is starting it. This is where *Great Jobs for Math Majors* comes in. Designed to help you put your major to work, this handy guide covers the basics of a job search and provides detailed profiles of careers in math. From the worlds of finance and science to manufacturing and education, you'll explore a variety of job options for math majors and determine the best fit for your personal, professional, and practical needs.

Do you want to be an actuary? Work in the banking industry? Program computers? In this updated edition, you'll find:

- Job-search basics such as crafting résumés and writing cover letters
- Self-assessment exercises to help determine your professional fit
- Investigative tools to help you find the perfect job
- True tales from practicing professionals about everyday life on the job
- Current statistics on earnings, advancement, and the future of the profession
- Resources for further information, including journals, professional associations, and online resources

STEPHEN E. LAMBERT is the director of career services at Plymouth State College in Plymouth, New Hampshire.

RUTH J. DeCOTIS is the director of recruitment services at Plymouth State College.

Teacher ▪ Actuary ▪ Mathematician ▪ Statistician
▪ Operations Research Analyst ▪ Marketing Research
▪ Financial Analyst ▪ Sales Representative
▪ Retail Buyer ▪ Purchasing Agent

ISBN 0-07-144859-4

The McGraw·Hill Companies
Visit us at: www.books.mcgraw-hill.com

$15.95 USA
$21.95 CAN

9 780071 448598

Cover design by Scott Rattray